OKLAHOMA
Total Eclipse Guide

Official Commemorative 2024 Keepsake Guidebook

2024 Total Eclipse State Guide Series

Aaron Linsdau

Sastrugi Press

Jackson Hole

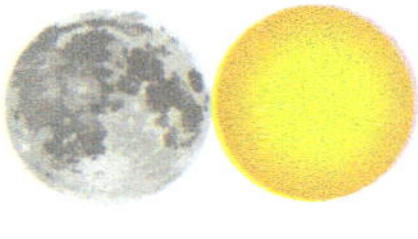

Sastrugi Press / Published by arrangement with the author

Oklahoma Total Eclipse Guide: Official Commemorative 2024 Keepsake Guidebook

The author has made every effort to accurately describe the locations contained in this work. Travel to some locations in this book is hazardous. The publisher has no control over and does not assume any responsibility for author or third-party websites or their content describing these locations, how to travel there, nor how to do it safely. Refer to local regulations and laws.

Any person exploring these locations is personally responsible for checking local conditions prior to departure. You are responsible for your own actions and decisions. The information contained in this work is based solely on the author's research at the time of publication and may not be accurate in the future. Neither the publisher nor the author assumes any liability for anyone climbing, exploring, visiting, or traveling to the locations described in this work. Climbing is dangerous by its nature. Any person engaging in mountain climbing is responsible for learning the proper techniques. The reader assumes all risks and accepts full responsibility for injuries, including death.

Sastrugi Press
PO Box 1297, Jackson, WY 83001, United States
www.sastrugipress.com
Quantity sales: Special discounts are available on quantity purchases by corporations, associations, and others. For details, contact the publisher at the address above.

Library of Congress Catalog-in-Publication Data
Library of Congress Control Number: 2017917785
Linsdau, Aaron
Oklahoma Total Eclipse Guide / Aaron Linsdau-1st United States edition
p. cm.
1. Nature 2. Astronomy 3. Travel 4. Photography
Summary: Learn everything you need to know about viewing, experiencing, and photographing the total eclipse in Oklahoma on April 8, 2024.

ISBN-13: 978-1-64922-314-2

508.4—dc23

All photography, maps and artwork by the author, except as noted.

10 9 8 7 6 5 4 3

Contents

Introduction 4

All About Oklahoma 6

Overview of Oklahoma 6

Weather 11

Forest Fires 13

Road Closures Due to Fires 14

Wilderness and Forest Park Safety 15

Eclipse Day Safety 17

All About Eclipses 19

Total vs Partial Eclipse 20

Early Myth & Astronomy 22

Contemporary American Solar Phenomena 24

Future American Eclipses 26

Viewing and Photographing the Eclipse 27

Planning Ahead 28

Understanding Sun Position 29

Eclipse Data for Selected Locations 31

Eclipse Photography 31

Eclipse Photography Gear 34

Camera Phones 36

Viewing Locations Around Oklahoma 43

Remember the Oklahoma Total Eclipse 65

Introduction

Thank you for purchasing this book. It has everything you need to know about the total eclipse in Oklahoma on April 8, 2024.

A total eclipse passing through the United States is a rare event. The last US total eclipse was in 2017. It traveled from Oregon to South Carolina. The last American total eclipse prior to that was in 1979!

The next total eclipse over the US will not be until April 8, 2024. It will pass over Texas, Oklahoma, the Midwest, and on to Maine. After that, the next coast-to-coast total eclipse will be in 2045.

It's imperative to make travel plans early. You will be amazed at the number of people swarming to the total eclipse path. Some might say watching a partial versus a total eclipse is a similar experience. It's not.

This book is written for Oklahoma visitors and anyone else viewing the eclipse. You will find general planning, viewing, and photography information inside. Should you travel to the eclipse path in Oklahoma in April, be prepared for an epic trip. The estimates based on the 2017 eclipse suggest that millions will converge on Oklahoma.

Some hotels in the communities and cities along the path of totality in Oklahoma have already been contacted by people to make reservations. Finding lodging along the eclipse path may be a major challenge.

Resources will be stretched far beyond the normal limits. Think gas lines from the late 1970s. It may be likely that traffic along highways will come to a complete standstill during this event. Be prepared with backup supplies.

Many smaller Oklahoma towns are far from any major city. Oklahoma country roads can be slow. Please obey posted speed limits for the safety of everyone. Be cautious about believing a map application's estimate of travel time in Oklahoma.

People in communities along the path of the total eclipse may rent out properties for this event. With this major celestial spectacle in the spring of 2024, be assured that Oklahoma "hasn't seen anything yet."

Is this to say to avoid Oklahoma or other areas during the eclipse? Not at all! This guidebook provides ideas for interesting, alternative, and memorable locations to see the eclipse. It will be too late to rush to a better spot once the eclipse begins. Law enforcement will be out to help drivers reconsider speeding.

Please be patient and careful. There will be a large rush of people from all over the world, converging on Oklahoma to enjoy the total eclipse. Be mindful of other drivers on eclipse weekend, as they may not be familiar with Oklahoma roads.

You should feel compelled to play hooky on April 8. Ask for the day off. Take your kids out of school. They'll be adults before the next chance to see a total eclipse over America. Create family memories that will last a lifetime. Sastrugi Press does not normally advocate skipping school or work. Make an exception because this is too big an event to miss.

Wherever you plan to be along the total eclipse path, leave early and remember your eclipse glasses. People from all around the planet will converge on Oklahoma. Be good to your fellow humans and be safe. We all want to enjoy this spectacular show.

Visit www.sastrugipress.com/eclipse for the latest updates for this state eclipse book series.

Author Information

Polar explorer and motivational speaker Aaron Linsdau's first book, *Antarctic Tears*, is an emotional journey into the heart of Antarctica. He ate two sticks of butter every day to survive. Aaron coughed up blood early in the expedition and struggled with equipment failures. Despite the endless difficulties, he set a world record for surviving the longest solo expedition to the South Pole.

Aaron teaches how to build resilience to overcome adversity by managing attitude. He shares his techniques for overcoming adrenaline burnout and constant overload. He inspires audiences to face their challenges with a new perspective. Aaron speaks about adversity, safety, and deailng with change. He hopes that you will be inspired and have an enjoyable time watching the total eclipse.

Visit his website at www.aaronlinsdau.com.

All About Oklahoma

Overview of Oklahoma

Nature lovers will find southeastern Oklahoma an ultimate paradise. It is a wonderful retreat from the hustle and bustle of city life. It offers various outdoor activities such as hiking, mountain climbing, biking, and nature immersion. If you want to experience the 2024 total eclipse in a unique location, this is one state that you can consider to enjoy a magnificent view.

Oklahoma is the forty-sixth state of the United States. The land it occupies was a portion of the Louisiana Purchase made in 1803. Situated in the South Central region of the United States, it is a major producer of multiple agricultural products, oil, and natural gas. The state is full of American-Indian culture that spans more than 30,000 years.

There are countless things to do and many places to see in Oklahoma. It holds some of the country's well-preserved museums and national historic sites. Aside from having a rich history and diverse culture, it has also been gifted with abundant natural resources. Also known as "Native America," Oklahoma is the home of Wichita Mountains National Wildlife Refuge. It is situated in the southwest area of the state, not too far from Lawton. It is referred to as the oldest wildlife facility of its kind in the United States. The famous Turner Falls, located in the Arbuckle Mountains, is another spectacular place to visit. It is a seventy-seven-foot-high waterfall that serves as a sanctuary for the people in Oklahoma to escape to nature. There are also caves and campsites, plus a lot of food to eat while in the area.

If you want ultimate relaxation and a calming atmosphere, Lake Wister State Park is the perfect retreat for you. It is situated in the

Ouachita Mountains and offers a magnificent backdrop of nature and trees. It is a peaceful and scenic place that is perfect for a soul-searching trip, for rest and relaxation, or for a romantic getaway. It offers a refreshing vibe suitable for those who want to immerse themselves in the beauty of nature. There are fifteen cabins in this park. Each is equipped with its own fireplace.

Another place to visit is the marvelous Broken Bow Lake. Famous for its blue waters and scenic location, Broken Bow Lake offers water activities such as scuba diving, boating, fishing, and swimming. It is perfect for those who love water sports. Aside from the exciting activities and great amenities, Broken Bow Lake also offers a picture-perfect and brilliant location to photograph the eclipse.

Family and wholesome fun can also be enjoyed at Hugo. It is the perfect place to have a picnic and to go fishing. You can enjoy the beach, go swimming, or simply relax while treating yourself to some fresh air. Kids can also play in the playground located at Raymond Gary State Park, which offers great amenities and activities for kids. The Hugo Lake State Park can accommodate groups and families of all sizes.

If food parks and festivals are more to your liking, Oklahoma can virtually fill the calendar with its sports events, cultural gatherings, and various festivals. The people in Oklahoma love parties and social gatherings. In fact, there is a festival almost every weekend. Some of the best ones are the Fourth of July fireworks display, weekend music festivals, local specialty events such as the Rush Springs, Watonga Cheese Festival, Watermelon Festival, and El Reno's Onion-Fried Burger Days. Oklahoma Food Truck Parks also offer an excellent opportunity for people to gather together and socialize.

The 2024 total eclipse over the state is a historical milestone that will bring travel and tourism in Oklahoma to a whole new level of experience. Visitors can expect the best food fests, an upbeat atmosphere, and an exciting vibe all over the state, especially in the southeastern part. The event will be a fusion of nature and people coming together to witness a spectacular and rare event.

Oklahoma is a delightful and wondrous place that has much to offer visitors. Aside from its natural beauty and friendly environment, it offers great opportunities for people to discover a different side of America, one that is more laid-back, relaxed, pleasant, and filled with people who love to socialize and enjoy life. Oklahoma is more than just a place, it's a one-of-a-kind experience for the 2024 total eclipse.

Hotels and Motels During the Eclipse

Once excitement of the total eclipse over Oklahoma spreads, rooms will become scarce. Many hotels in towns along the path of totality in western states sold out for a year or more during the 2017 total eclipse. Oklahoma is not alone in this challenge. Hotels all along the path of totality will sell out in anticipation of the 2024 total eclipse.

What does this mean for eclipse visitors? Lodging and room rentals in eclipse towns will be at a massive premium. Does that mean all hope is lost to find a place to stay? Not at all. But you will have to be creative. There will be few, if any, hotel rooms available in these eclipse cities by early 2024. Accommodations in the cities and towns along the path of the eclipse will be difficult to come by.

In summer 2017, the author searched on Hotels.com for rooms along the 2017 total eclipse path on the weekend of August 21 and found many major cities sold out. Once word of the 2024 eclipse spreads, room rates will increase and availability will drop.

Search for rooms farther away from the eclipse path. If you are willing to stay in cities outside the eclipse path, you will have better success at finding rooms. As the eclipse approaches, people will book rooms farther from the totality path. By early spring, rooms in cities near the total eclipse path may be unavailable. The effect of this event will be felt across Oklahoma and the rest of the United States.

Think regionally when looking for rooms. Be prepared to search far and wide during this major event. If a five-hour drive is manageable, your lodging options greatly expand, but it also increases your travel risk.

Internet Rentals

To find rooms to stay in towns along the eclipse path, try a web service such as Airbnb.com. Note that some people rent out rooms or homes illegally, against zoning regulations. Cities will feel the crunch of inquiries early due to others who experienced the 2017 eclipse.

If cities fully enforce zoning laws, authorities may prevent your weekend home rental. Online home rentals during the eclipse will be a target for rental scams. People from out of the area steal photos and descriptions, then post the home for rent. You send your check or wire money to a “rental agent” then show up to find you have been scammed. If the deal sounds strange or too good to be true, run away.

Camping

If you can book a campsite, do it as soon as you can. Do not wait. All areas in the national forests are first-come, first-served. Forest roads may be packed. Expect all areas to be swarming with people. Show up early to stake out your spot. Consider staying farther away and driving early on April 8.

Please respect private land too. Oklahoma folks don’t take kindly to people overrunning their property without permission. In a big state with millions of residents, people are very protective, but they’re friendly, too. You never know what you might be able to arrange with a smile and a bit of money.

This all said, there are plenty of camping opportunities throughout Oklahoma. You don’t have to sleep exactly on the eclipse path. If you’re ready to rough it, there are national forest camping options.

Government agencies will meet years in advance to talk about how to manage the influx of people. Every possible government agency will be working full time to enforce the various rules and regulations.

National Parks and Monuments

Finding a camping site at any state park, national park, or national monument along the eclipse path in Oklahoma will be challenging. To watch the eclipse from any location, you do not have to sleep in it. You just need to drive there in the morning.

Law enforcement will be present on the eclipse weekend. Hundreds of thousands of people are expected in the region. Parking may overflow. It will make parking lots and lines on Black Friday at the mall look uncrowded. For an event of this magnitude, find your location as early as possible.

The first sentence of the national parks mission statement is:

"The National Park Service preserves unimpaired the natural and cultural resources and values of the national park system for the enjoyment, education, and inspiration of this and future generations."

Roadside camping (sleeping in your car) is not allowed in national monuments or parks. Park facilities are only designed to handle so many people per day. Water, trash collection, and toilets can only withstand so much. If you notice trash on the ground, take a moment to throw it away. Protect your national park and help out. Rangers are diligent and hardworking but they can only do so much to manage the expected crowds.

National Forests and Wilderness

There are national forest options in Oklahoma. They all have camping opportunities. The forest service manages undeveloped and primitive campsites. Be sure to check for any fire restrictions. Check with individual agencies for last-minute information and regulations. The forest service requires proper food storage. Plan to purchase food and water before choosing your campsite. Below is a partial list of national forests along or near the total eclipse path:

Ouachita NF:
https://www.fs.usda.gov/ouachita

Ozark NF:
https://www.fs.usda.gov/osfnf

Backcountry service roads abound in Oklahoma. Maps for forests are available at local visitor centers and bookstores. This book's website has digital copies of some forest maps.

Printed national forest maps are large and detailed. They have illustrated road paths, connections, and other vital travel information not available on digital device maps. Viewing digital maps on your smartphone or mobile pad is difficult. If you plan to camp in the forest, a real paper map is a wise investment.

Camping in federal wilderness areas is also allowed. Those areas afford the ultimate backcountry experience. However, be aware that no vehicle travel is allowed in these specially designated areas. This ban includes: vehicles, bikes, hang gliders, and drones. You can travel only on foot or with pack animals.

Sleep in Your Car

Countless RVs, campers, trucks, cars, and motorcycles will flood Oklahoma. Sleeping in your car with friends is tolerable. Doing so with unadventurous spouses or children is another matter.

Do not be caught along the path of the total eclipse without some sort of plan, especially in the bigger cities of Oklahoma. The whole path of totality will fill with people on April 8.

Useful Local Webcams

Local webcams are handy to make last-minute travel decisions. Modern webcams are sensitive enough to show headlights at night. Use them to determine if there are issues before traveling out. Eclipse traffic will add to the morning commuter traffic.

There are smartphone applications which are useful to check webcams in many locations. Consult your device's app store for the latest updates. Whether you use an app or computer, an Internet search will reveal many handy webcams for your eclipse planning.

Weather

It's all about the weather during the eclipse. Nothing else will matter if the sky is cloudy. You can be nearly anywhere along the path in Oklahoma and catch a view of the event when traffic comes to a standstill. But if there's a cloud cover forecast, seriously reconsider your viewing location.

Travel early wherever you plan to go. Attempting to change locations an hour before the eclipse due to weather will likely cause you to miss the event. Oklahoma country roads can be narrow and slow. The number of vehicles will cause unexpected backups.

Modern Forecasts

Use a smartphone application to check the up-to-date weather. Wunderground is a good application and has relatively reliable forecasts for the region. The hourly forecast for the same day has been rather accurate for the last two years. The below discussion refers to features found in the Wunderground app. However, any application with detailed weather views will improve your eclipse forecasting skills.

Cloud Cover Forecast

The most useful forecast view is the visible and infrared cloud-coverage map. Avoid downloading this app the night before and trying to learn how to read it. Practice reading them at home. It's imperative to understand how to interpret the maps early.

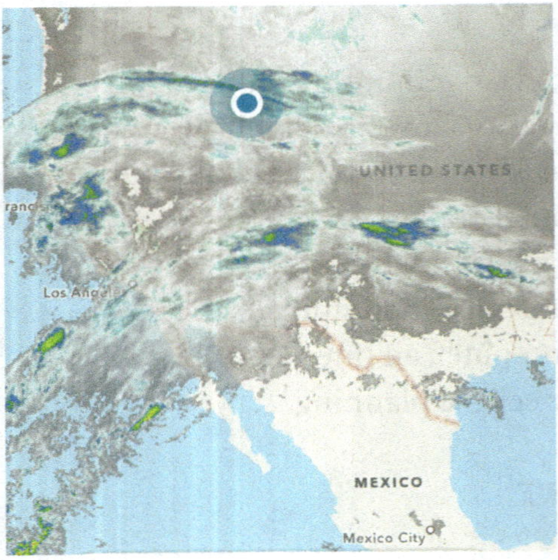

Infrared cloud map showing the worst case eclipse cloud cover. Courtesy of National Weather Service.

All cloud cover, night or day, will appear on an infrared map. Warm, low-altitude clouds are shown in white and gray. High-altitude cold clouds are displayed in shades of green, yellow, red, and purple. Anything other than a clear map spells eclipse-viewing problems.

To improve your weather guess, use the animated viewer of the cloud cover. It will give you a sense of cloud motion. You can discern whether clouds or rain are moving toward, away from, or circulating around your location.

Normal Oklahoma Weather Pattern

Due to the direction of the jet stream, most weather travels across the Pacific Ocean, through the western states, over the Rockies, and then into Oklahoma. On occasion, weather can approach from Mexico or off the Gulf of Mexico. Due to the nature of the tropical storms from the Atlantic, weather in Oklahoma can be unpredictable.

The common weather pattern in April is slightly warm in the afternoon and mildly cool in the evenings. Passing cold fronts in spring can bring unexpected cloud cover and rains.

Historically, Oklahoma tends to have moderate cloud cover during April. Prepare to make adjustments. If anything other than clear skies are predicted, drive to other parts of Texas, Arkansas, or Oklahoma.

Be aware of tornadoes in Oklahoma. Although the peak tornado season is June, there have been many recorded tornadoes in April. Pay attention to the weather forecast. If dangerous weather is predicted, your main concern should be safety rather than chasing an eclipse.

Consider that slow-moving clouds can obscure the sun for far longer than the four-minute duration of the totality. The time of totality is so short that you do not want to risk it. Missing it due to a single cloud will be a major disappointment.

Local Eclipse Weather Forecasts

Local town and city newspapers, radio, and television stations around Oklahoma will have a weekend edition with articles discussing the eclipse weather. However, conditions change unpredictably in Oklahoma. A three-day forecast may be incorrect.

Forest Fires

For the past several years, forest fires have been common in the United States. The spring of 2024 will likely to be no different. There have been fires in the Oklahoma during spring. With the ongoing drought, there will possibly be fires again during the time of the eclipse. For fire updates check:

inciweb.nwcg.gov

For the best eclipse viewing experience, you need to have as clear a sky as possible. Fog, clouds, or smoke will obscure the subtleties of the sun's corona. If you think the view of the sky is going to be blocked, don't wait until the last minute to move to a clearer location. If you wait too long to decide to move to a better viewing area, it may be impossible due to traffic.

Road Closures Due to Fires

Highways connecting various Oklahoma towns can be closed during wildfires. During April 2016, there were multiple large fire incidents including the Fish Creek, Pittsburg South, and Salt Hollow Fires.

With unpredictable weather in the last few years, it's a guess what will happen in April 2024. If forest areas continue to remain dry, the whole region may have fires.

Although fire is an important part of forest ecology, it worries eclipse chasers. Other than clouds, smoke from fires will block the view of the sun and moon on the morning of the eclipse. Should there be fires where you are or may be headed, reconsider your location as early as possible. The most accurate website for fires is:

inciweb.nwcg.gov

Check the Oklahoma road report for updated information:

ok.gov/odot/

It's imperative to plan for fires and their effects. Watch the weather reports. If strong winds and lightning storms are forecast, prepare to change your viewing location. If conditions are poor, you and thousands of other vehicles will be trapped in slow-moving traffic.

If you believe it's necessary to leave a town to watch the eclipse, do so the night before or extremely early in the morning. RVs are common, and trains of them crawl through popular areas.

Oklahoma Information

Cellular Phones

Cellular "cell" phone service in remote Oklahoma locations may be problematic. Most of the time there is good coverage along the main highways and interstates. However, even along major thoroughfares, there can be little or no coverage.

It's possible to find zones where text messages will send when phone calls are impossible. If you cannot make a phone call, the chance of having data coverage for web surfing or e-mail is low.

Please look up any information or communicate what you need before departing from the main roads around Oklahoma. Bureau of Land Management (BLM) areas sometimes have coverage. Planned to be self-contained. Plan for your cell phone not to connect.

You may find yourself out of cell service. With a large number of cell users in a concentrated area, coverage and data speed may collapse as well. Search on the phrase "cell phone coverage breathing".

Wilderness and Forest Safety

All Oklahoma mountains and wilderness areas are full of wild animals. Although beautiful, wild animals can be dangerous. They can easily injure or kill people, as they are far more powerful than humans. Do not try to feed any wild animals, including squirrels, foxes, and chipmunks, as they can carry diseases. These suggestions apply to all public lands.

Feral Hogs

Introduced in the early twentieth century, hogs have become a major problem in Oklahoma. Although they tend to flee when encountered, hogs have been known to attack people. They can be aggressive and their tusks can inflict serious wounds. There have been instances of deaths, too. It is best to leave these animals alone if you encounter one.

Alligators

Although the chance of encountering an alligator is low, it is not uncommon to encounter them. With hurricanes moving significant quantities of water in recent years, alligators may be in unexpected locations. If you encounter one, back away from the animal and contact fish and game if it poses a threat.

Venomous Snakes

There are multiple species of venomous snakes in Oklahoma including the Western Diamondback & Timber Rattlesnakes, Pygmy, Copperhead, Coral Snake, and Cottonmouth. Although these reptiles are not generally aggressive, they can strike when provoked or threatened. Of the approximately 8,000 people annually bitten by venomous snakes in the United States, ten to fifteen people die according to the U.S. Food and Drug Administration.

The best way to avoid rattlesnake encounters is to be mindful of your environment. Do not place your hands or feet in locations where you cannot clearly see the surroundings. Avoid heavy brush or tall weeds where snakes hide during the day. Step on a log or rock rather than over it, as a hidden snake might be on the other side. Rattlesnakes may not make any noise before striking.

Avoid handling all snakes. Should you be bitten, stay calm and call 911 or emergency dispatch as soon as possible. Transport the victim to the nearest medical facility immediately. Rapid professional treatment is the best way to manage rattlesnake bites. Refer to US Forest Service and professional medical texts for more information on managing rattlesnakes injuries.

Bears

The forests of Oklahoma are home to black bears. Safety is imperative around these powerful animals. Although they often appear docile, they can become aggressive if threatened. In the unlikely event of an attack, fight back against the bear. Use whatever you have at your disposal to defend yourself. Report aggressive bear activity to the local authorities.

If a bear hears you, it will usually vacate the area. Bear charges are often caused by unexpected and surprise encounters. Noise is the best defense to avoid surprising bears. Regularly clap, make noise, and talk loudly. The Oklahoma Department of Wildlife Conservation website has more specific information on safety and food management in bear country at www.wildlifedepartment.com/media/bear.htm.

It is recommended to stay one hundred yards (300 feet) away from all bears. They are exciting to see but need their space. Refer to current forest or park regulations for more safety information.

Mountain Lions

Though listed as rarely seen in the state, there have been recent reports of mountain lions in Oklahoma by the Oklahoma Department of Wildlife Conservation. If you encounter a mountain lion, do not run. Keep calm, back away slowly, and maintain eye contact. Do all you can to appear larger. Stand upright, raise your arms, or hoist your jacket. Never bend over or crouch down. If attacked, fight back.

Eclipse Day Safety

1. Hydrate

Spring temperatures are usually mild to warm. The excitement of the event can distract you from managing hydration. Drink plenty of water. Consume more than you would at home.

2. Eye Safety time

Use certified eclipse safety glasses at all times when viewing the partial eclipse. Only remove the glasses when the totality happens. Give your eyes time to rest. They can dry out and become irritated. Bring FDA approved eye drops to keep your eyes moist.

3. Sun exposure

Facing at the sun for three hours can result in sunburns. Wear sunglasses and liberally apply sunscreen to avoid sunburns.

4. Eat well

Keep your energy up. Appetite loss is common when traveling. Maintain your normal eating schedule.

5. Prepare for temperature changes

Temperatures will drop rapidly during the eclipse and also once the sun sets. Bring appropriate clothing.

6. Talk with your doctor

If the humidity or heat bothers you talk with your doctor before traveling. Seek professional medical attention for serious symptoms.

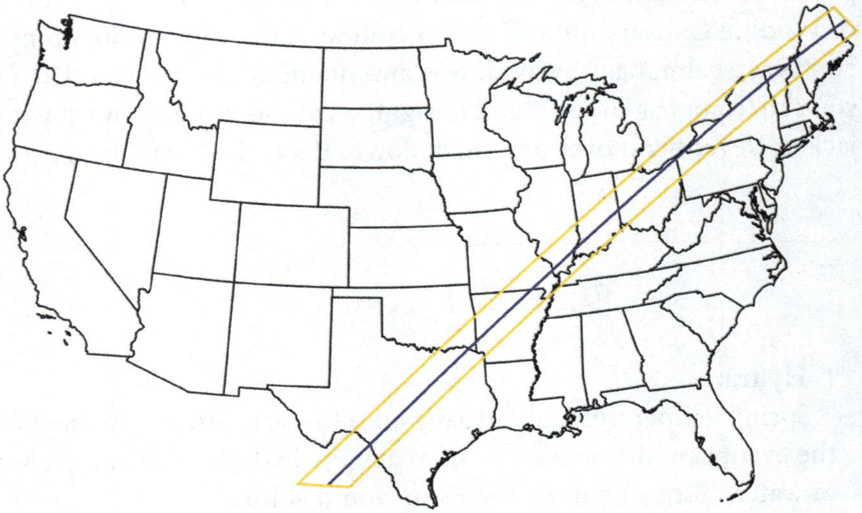

Total eclipse path across the United States (approximate).

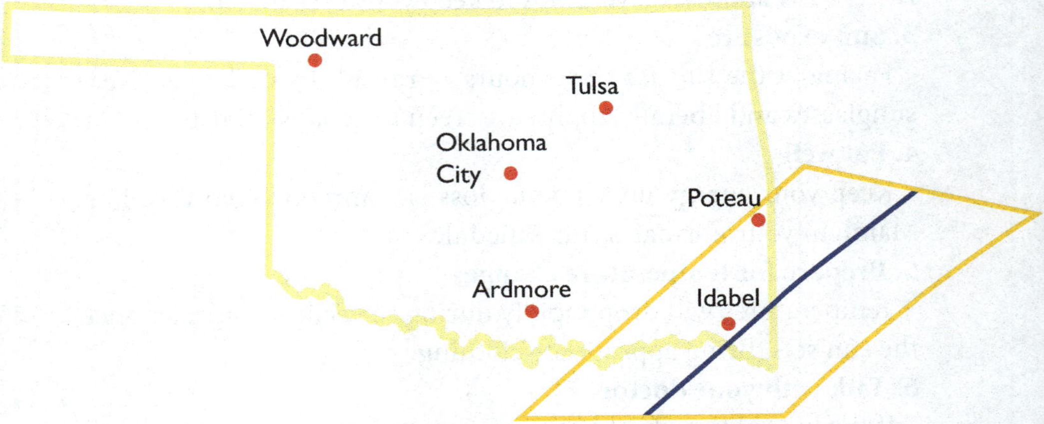

Total eclipse path across Oklahoma (approximate).

All About Eclipses

How an Eclipse Happens

An eclipse occurs when one celestial body falls in line with another, thus obscuring the sun from view. This occurs much more often than you'd think, considering how many bodies there are in the solar system. For instance, there are over 150 moons in the solar system. On Earth, we have two primary celestial bodies: the sun and the moon. The entire solar system is constantly in motion, with planets orbiting the sun and moons orbiting the planets. These celestial bodies often come into alignment. When these alignments cause the sun to be blocked, it is called an eclipse.

For an eclipse to occur, the sun, Earth, and moon must be in alignment. There are two types of eclipses: solar and lunar. A solar eclipse occurs when the moon obscures the sun. A lunar eclipse occurs when the moon passes through Earth's shadow. Solar eclipses are much more common, as we experience an average of 240 solar eclipses a century compared to an average of 150 lunar eclipses. Despite this, we are more likely to see a lunar eclipse than a solar eclipse. This is due to the visibility of each.

For a solar eclipse to be visible, you have to be in the moon's shadow. The problem with viewing a total eclipse is that the moon casts a small shadow over the world at any given time. You have to be in

* ILLUSTRATION NOT TO SCALE

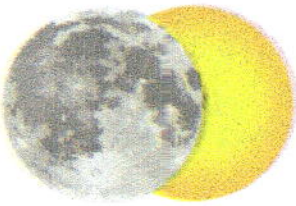

a precise location to view a total eclipse. The issue that arises is that most of these locations are inaccessible to most people. Though many would like to see a total solar eclipse, most aren't about to set sail for the middle of the Pacific Ocean. In fact, a solar eclipse is visible in the same place on the world on average every 375 years. This means that if you miss a solar eclipse above your hometown, you're not going to see another one unless you travel or move.

It's much easier to catch a glimpse of a lunar eclipse, even though they occur at a much lower frequency than their solar counterparts. A lunar eclipse darkens the moon for a few hours. This is different than a new moon when it faces away from the sun. During these eclipses, the moon fades and becomes nearly invisible.

Another result of a lunar eclipse is a blood moon. Earth's atmosphere bends a small amount of sunlight onto the moon turning it orange-red. The blood moon is caused by the dawn or dusk light being refracted onto the moon during an eclipse.

Lunar eclipses are much easier to see. Even when the moon is in the shadow of Earth, it's still visible throughout the world because of how much smaller it is than Earth.

Total vs. Partial Eclipse

What is the difference between a partial and total eclipse? A total eclipse of either the sun or the moon will occur only when the sun, Earth, and the moon are aligned in a perfectly straight line. This ensures that either the sun or the moon is partially or completely obscured.

In contrast, a partial eclipse occurs when the alignment of the three celestial bodies is not in a perfectly straight line. These types of eclipses usually result in only a part of either the sun or the moon being obscured. This is often what led to ancient civilizations believing that some form of magical beast or deity was eating the sun or the moon. It appears as though something has taken a bite out of either the sun or the moon during a partial eclipse.

Total eclipses, rarer than partial eclipses, still occur quite often. It's more difficult for people to be in a position to experience such an event firsthand. Total solar eclipses can only be viewed from a small portion of the world that falls into the darkest part of the moon's shadow. Often this happens in the middle of the ocean.

The Moon's Shadow

The moon's shadow is divided into two parts: the umbra and the penumbra. The former is much smaller than the latter, as the umbra is the innermost and darkest part of the shadow. The umbra is thus the central point of the moon's shadow, meaning that it is extremely small in comparison to the entire shadow. For a total solar eclipse to be visible, you need to be directly beneath the umbra of the moon's shadow. This is because that is the only point at which the moon completely blocks the view of the sun.

In contrast, the penumbra is the region of the moon's shadow in which only a portion of the light cast by the sun is obscured. When

Total eclipse shadow 2016 as seen from 1 million miles on the Deep Space Climate Observatory satellite. Courtesy of NASA.

standing in the penumbra, you are viewing the eclipse at an angle. In the penumbra, the moon does not completely block the sun from view. This means that while the event is a total solar eclipse, you'll only see a partial eclipse. The umbra for the April 8 eclipse is over one hundred miles wide. The penumbra will cover much of the United States.

To provide some context, one total solar eclipse we experienced occurred on March 9, 2016, and was visible as a partial eclipse across most of the Pacific Ocean, parts of Asia, and Australia. However, the only place in the world to view this total solar eclipse was in a few parts of Indonesia.

Due to the varied locations and the brief periods for which they're visible, it's difficult to see each and every eclipse that occurs. The umbra of the moon is such a small fraction of the entire shadow and the majority of our planet is comprised of water. Thus, the rarity of being able to view a total solar eclipse increases significantly because it's likely that the umbra will fall over some part of the ocean rather than a populated landmass. There are not total or annular eclipses every month because the moon's orbit is 5.1° off the ecliptic plane of the Earth and sun.

Eclipses Throughout History

Ancient peoples believed eclipses were from the wrath of angry gods, portents of doom and misfortune, or wars between celestial beings. Eclipses have played many roles in cultures, creating myths since the dawn of time. Both solar and lunar eclipses affected societies worldwide. Inspiring fear, curiosity, and the creation of legends, eclipses have cast a long shadow in the collective unconscious of humanity throughout history.

Early Myth & Astronomy

Documented observations of solar eclipses have been found as far back in history as ancient Egyptian and Chinese records. Timekeeping was important to ancient Chinese cultures. Astronomical observations were an integral factor in the Chinese calendar. The first observation of a solar eclipse is found in Chinese records from over 4,000 years ago. Evidence suggests that ancient Egyptian observations may predate those archaic writings.

Many ancient societies, including Roman, Greek and Chinese civilizations, were able to infer and foresee solar eclipses from astronomical data. The sudden and unpredictable nature of solar eclipses had a stressful and intimidating effect on many societies that lacked the scientific insight to accurately predict astronomical events. Relying on the sun for their agricultural livelihood, those societies interpreted solar eclipses as world-threatening disasters.

In ancient Vietnam, solar eclipses were explained as a giant frog eating the sun. The peasantry of ancient Greece believed that an eclipse was the sign of a furious godhead, presenting an omen of wrathful retribution in the form of natural disasters. Other cultures were less speculative in their investigations. The Chinese Song Dynasty scientist Shen Kuo proved the spherical nature of the Earth and heavenly bodies through scientific insight gained by the study of eclipses.

The Eclipse in Native American Mythology

Eclipses have played a significant role in the history of the United States. Before Europeans settled in the Americas, solar eclipses were important astronomical events to Native American cultures. In most native cultures, an eclipse was a particularly bad omen. Both the sun and the moon were regarded as sacred. Viewing an eclipse, or even being outside for the duration of the event, was considered highly taboo by the Navajo culture. During an eclipse, men and women would simply avert their eyes from the sky, acting as though it was not happening.

The Choctaw people had a unique story to explain solar eclipses. Considering the event as the mischievous actions of a black squirrel and its attempt to eat the sun, the Choctaw people would do their best to scare away the cosmic squirrel by making as much noise as possible until the end of the event, at which point cognitive bias would cause them to believe they'd once again averted disaster on an interplanetary scale.

Contemporary American Solar Phenomena

The investigation of solar phenomena in twentieth-century American history had a similarly profound effect on the people of the United States. A total solar eclipse occurring on the sixteenth of June, 1806, engulfed the entire country. It started near modern-day Arizona. It passed across the Midwest, over Ohio, Pennsylvania, New York, Massachusetts, and Connecticut. The 1806 total eclipse was notable for being one of the first publicly advertised solar events. The public was informed beforehand of the astronomical curiosity through a pamphlet written by Andrew Newell entitled *Darkness at Noon, or the Great Solar Eclipse*.

This pamphlet described local circumstances and went into great detail explaining the true nature of the phenomenon, dispelling myth and superstition, and even giving questionable advice on the best methods of viewing the sun during the event. Replete with a short historical record of eclipses through the ages, the *Darkness at Noon* pamphlet is one of the first examples of an attempt to capitalize on the mysterious nature of solar eclipses.

Another notable American solar eclipse occurred on June 8, 1918. Passing over the United States from Washington to Florida, the eclipse was accurately predicted by the U.S. Naval Observatory and heavily documented in the newspapers of the day. Howard Russell Butler, painter and founder of the American Fine Arts Society, painted the eclipse from the U.S. Naval Observatory, immortalizing the event in *The Oregon Eclipse*.

Four more total solar eclipses occurred over the United States in the years 1923, 1925, 1932, and 1954, with another occurring in 1959. The October 2, 1959, solar eclipse began over Boston, Massachusetts. It was a sunrise event that was unviewable from the ground level. Eminent astronomer Jay Pasachoff attributed this event to sparking his interest in the study of astronomy. Studying under Professor Donald Menzel of Williams College, Pasachoff was able to view the event from an airline hired by his professor.

To this day, many myths surround the eclipse. In India, some local customs require fasting. In eastern Africa, eclipses are seen as a danger to pregnant women and young children. Despite the mystery and legend associated with unique and rare astronomical events, eclipses continue to be awe-inspiring. Even in the modern day, eclipses draw out reverential respect for the inexorable passing of celestial bodies. They are a reminder of the intimate relationship between the denizens of Earth and the universe at large.

Present Day Eclipses

The year 2017 brought the world's most-watched total eclipse in history on August 21, when a total solar eclipse crossed the United States. An annular eclipse, a "ring of fire," will pass over the United States in 2023 from Oregon to Texas. Though impressive, it will not

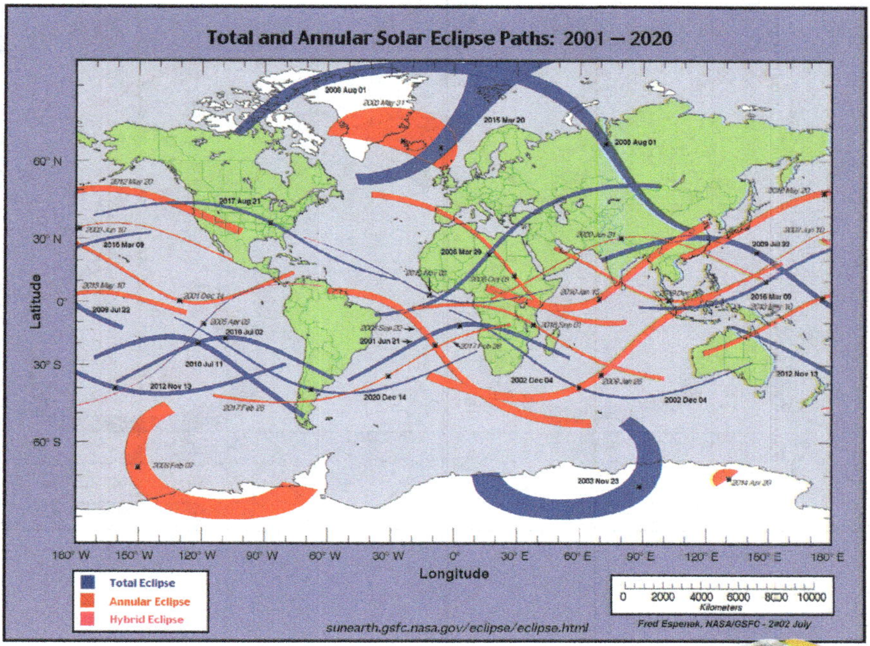

COURTESY OF NASA

compare to the 2024 total eclipse. There is little in nature that equals the spectacle of the sun's corona and seeing stars in the day.

There will be multiple partial, annular, or hybrid eclipses across the world before the 2024 total eclipse. However many are in remote, inaccessible, or potentially dangerous locations on the globe. In 2019 and 2020, Chile and Argentina will experience total eclipses. The next total eclipse after that will occur over Antarctica in 2021. An extremely rare hybrid eclipse will happen in 2023 over the Indian Ocean, Australia, and Indonesia.

The next total solar eclipse viewable from the United States will occur on April 8, 2024. It will be visible in fourteen states: Texas, Oklahoma, Arkansas, Missouri, Kentucky, Illinois, Indiana, Ohio, Pennsylvania, Michigan, New York, Vermont, New Hampshire, and Maine.

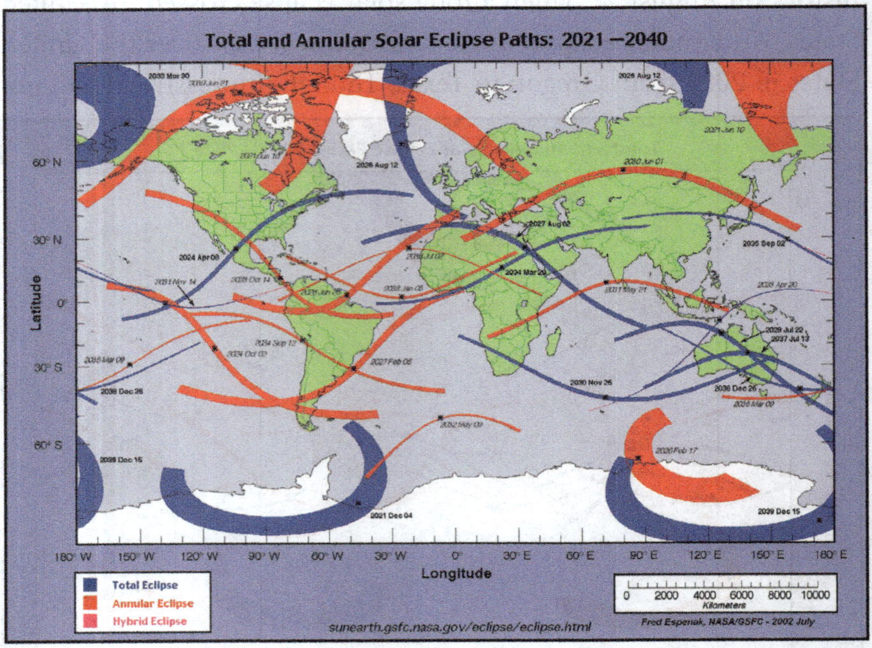

COURTESY OF NASA

Viewing and Photographing the Eclipse

At-home Pinhole Method

Use the pinhole method to view the eclipse safely. It costs little but is the safest technique there is. Take a stiff piece of single-layer cardboard and punch a clean pinhole. Let the sun shine through the pinhole onto another piece of cardboard. That's it!

Never look at the sun through the pinhole. Your back should be toward the sun to protect your eyes. To brighten the image, simply move the back piece of cardboard closer to the pinhole. To see it larger, move the back cardboard farther away. Do not make the pinhole larger. It will only distort the crescent sun.

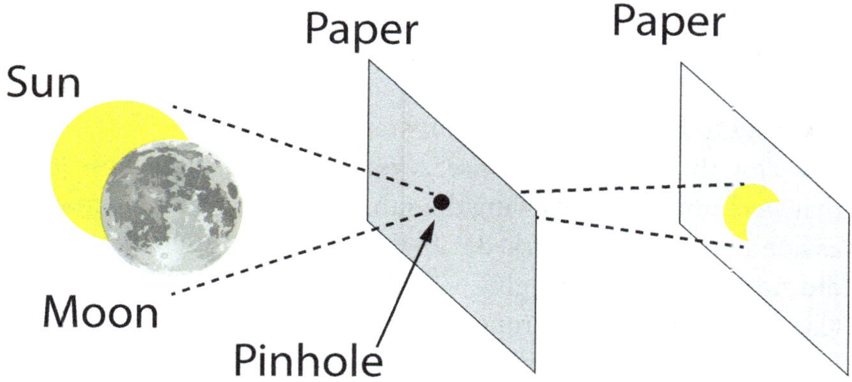

Welding Goggles

Welding goggles that have a rating of fourteen or higher are another useful eclipse viewing tool. The goggles can be used to view the solar eclipse directly. Do not use the goggles to look through binoculars or telescopes, as the goggles could potentially shatter due to intense direct heat. Avoid long periods of gazing with the goggles. Look away every so often. Give your eyes a break.

Solar Filters for Telescopes

The ONLY safe way to view solar eclipses using telescopes or binoculars is to use solar filters. The filters are coated with metal

to diminish the full intensity of the sun. Although the filters can be expensive, it is better to purchase a quality filter rather than an inexpensive one that could shatter or melt from the heat.

The filters attach to the front of the telescope for easy viewing. Remember to give your telescope cooling breaks. Rapid heating can damage your equipment with or without filters attached.

Watch Out for Unsafe Filters

There are several myths surrounding solar filters for eclipse viewing. In order for filters to be safe, they must be specially designed for looking at a solar eclipse. The following are all unsafe for eclipse viewing and can lead to retinal damage: developed colored or chromogenic film, black-and-white negatives such as X-rays, CDs with aluminum, smoked glass, floppy disk covers, black-and-white film with no silver, sunglasses, or polarizing films.

Watch Out for Unsafe Eclipse Glasses

During the 2017 total eclipse, several vendors sold eclipse glasses that were not safe for viewing the sun. Although they were marketed as safe and were even marked with the ISO 12312-2 certification, they did not block eye-damaging visible, infrared, and ultraviolet light. Check the American Astronomical Society's website (eclipse.aas.org) for a list of reputable eclipse glasses vendors.

Viewing with Binoculars

When viewing the eclipse with binoculars, it is important to use solar filters on both lenses until totality. Only then is it safe to remove the filter. As the sun becomes visible after totality, replace the filters for safe viewing. Protect your pupils. Remember to give your binoculars a cool-down break between viewings. They can overheat rapidly from being pointed directly at the sun even with filters attached.

Planning Ahead

There are many things to keep in mind when viewing a total eclipse. It is important to plan ahead to get the most out of this extraordinary experience.

Understanding Sun Position

All compass bearings in this book are true north. All compasses point to Earth's magnetic north. The difference between these two measurements is called magnetic declination. The magnetic declination for Okahoma is:

3° 35' E ± 0° 21' (for Oklahoma City in 2020)

Adjust the declination from the azimuth bearing as given in the text, and set your compass to that direction.

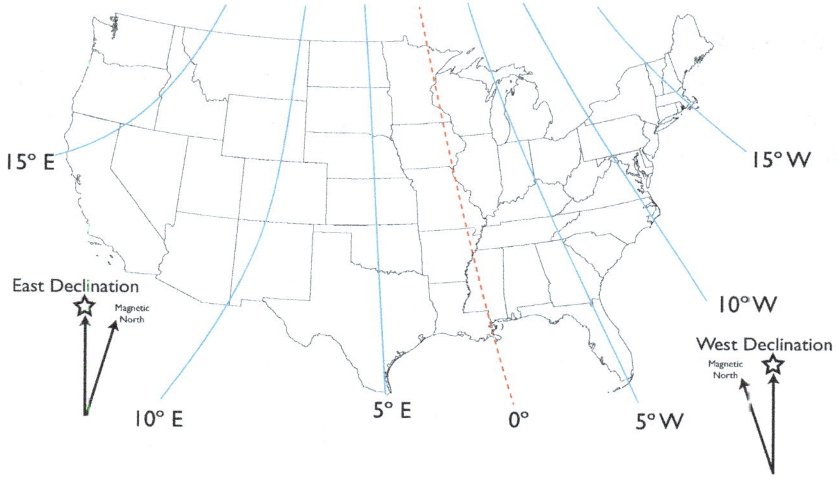

If you purchase a compass with a built-in declination adjustment, you can change the setting once and eliminate the calculations. The Suunto M-3G compass has this correction. A compass with a sighting mirror or wire will help you make a more accurate azimuth sighting.

The Suunto M-3G also has an inclinometer. This allows you to measure the elevation of any object above the horizon. Use this to figure out how high the sun will be above your position.

You can also use a smartphone inclinometer and compass for this purpose. Make sure to calibrate your smartphone's compass before every use, otherwise it might indicate the wrong bearing. Set the smartphone compass for true north to match the book. Understand the compass prior to April 8. There will be little time to guess or search on Google. Smartphone and GPS compasses are "sticky." Their compasses don't swing as freely as a magnetic compass does.

The author has used his magnetic compass for azimuth measurements and a smartphone to measure elevation. Combining these two tools will allow you to make the best sightings possible.

Outdoor sporting goods stores in most towns and cities carry compasses. Purchase and practice with a good compass in your hometown well before the event. Take the time to learn how to use it before the day of the eclipse. You do not want to struggle with orienteering basics under pressure.

Sun Azimuth

Azimuth is the compass angle along the horizon, with 0° corresponding to north, and increasing in a clockwise direction. 90° is east, 180° is south, and 270° is west.

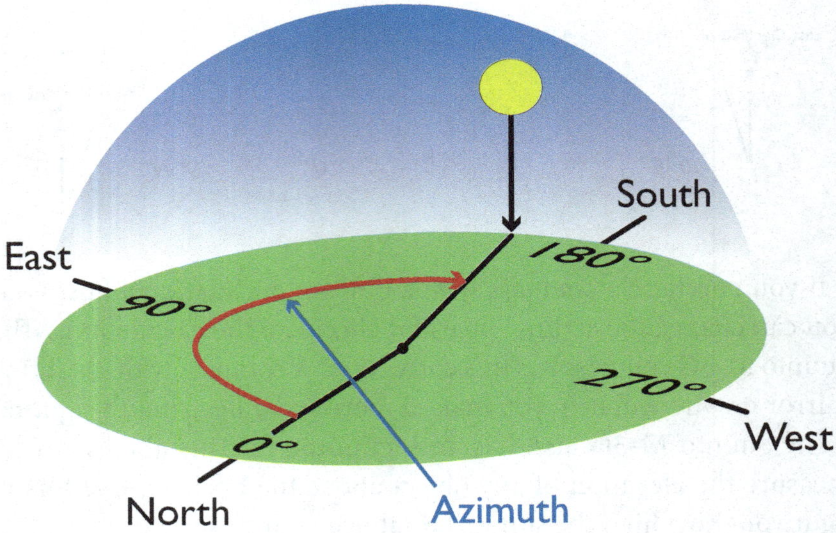

Sun Elevation

Altitude is the sun's angle up from the horizon. A 0° altitude means exactly on the horizon and 90° means "straight up."

Using the sun azimuth and elevation data, you can predict the position of the sun at any given time. Positions given in this book coincide with the time of eclipse totality unless otherwise noted.

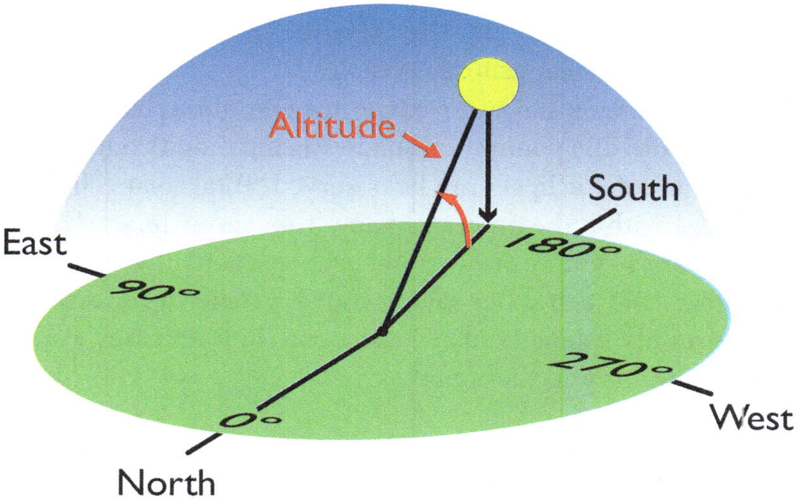

Eclipse Data for Select Oklahoma Locations

LOCATION	TOTALITY START (CDT)
ANTLERS	1:45:43PM
HONOBIA	1:46:42PM
HUGO	1:44:51PM
IDABEL	1:45:17PM
POTEAU	1:48:44PM
TALIHINA	1:47:46PM

Eclipse Photography

Photographing an eclipse is an exciting challenge, as the moon's shadow moves near 1,600MPH. There is an element of danger and the pressure of time. Looking at the unfiltered sun through a camera can permanently damage your vision and your camera. If you are unsure, just enjoy the eclipse with specially designed eclipse glasses. Keep a solar filter on your lens during the eclipse and remove for the duration of totality!

Partial Vs. Total Solar Eclipse

To successfully and safely photograph a partial and total eclipse, it is important to understand the difference between the two. A solar eclipse occurs when the moon is positioned between the sun and Earth. The region where the shadow of the moon falls upon Earth's surface is where a solar eclipse is visible.

The moon's shadow has two parts—the penumbral shadow and the umbral shadow. The penumbral shadow is the moon's outer shadow where partial solar eclipses can be observed. Total solar eclipses can only be seen within the umbral shadow, the moon's inner shadow.

You cannot say you've seen a total eclipse when all you saw was a partial solar eclipse. It is like saying you've watched a concert, but in reality, you only listened outside the arena. In both cases, you have missed the drama and the action.

Photographing A Partial And Total Solar Eclipse

Aside from the region where the outer shadow of the moon is cast, a partial solar eclipse is also visible before a total solar eclipse within

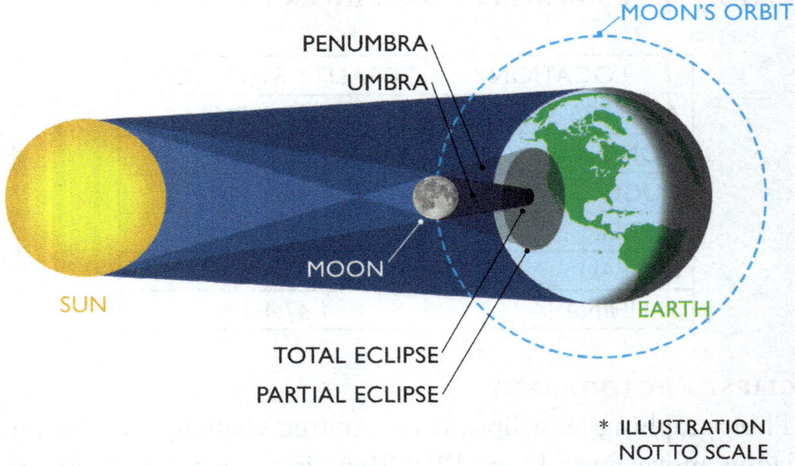

the inner shadow region. In both cases, it is imperative to use a solar filter on the lens for both photography and safety reasons. This is the only difference between taking a partial eclipse and a total eclipse photograph of the sun.

To photograph a total solar eclipse, you must be within the Path of Totality, the surface of the Earth within the moon's umbral shadow.

The Challenge

A total solar eclipse only lasts for a couple of minutes. It is brief, but the scenario it brings is unforgettable. Seeing the radiant sun slowly being covered by darkness gives the spectator a high level of anticipation and indescribable excitement. Once the moon completely covers the sun's radiance, the corona is finally visible. In the darkness, the sun's corona shines, capturing the crowd's full attention. Watching this phenomenon is a breathtaking experience.

Amidst all the noise, cheering, and excitement, you have no more than a few minutes to take a perfect photograph. The key to this is planning. You need to plan, practice, and perfect what you will do when the big moment arrives because there is no replay. The pressure is enormous. You only have a short time to capture the totality and the sun's corona using different exposures.

Plan, Practice, Perfect

It is important to practice photographing before the actual phenomenon arrives. Test your chosen imaging setup for flaws. Rehearse over and over until your body remembers what you will do from the moment you arrive at your chosen spot to the moment you pack up and leave the area.

You will discover potential problems regarding vibrations and focus that you can address immediately. This minimizes the variables that might affect your photographs at the most critical moment.

It's common for experienced eclipse chasers to lose track of what they plan to do. Write down what you expect to do. Practice it time and again. Play annoying, distracting music while you practice. Try photographing in the worst weather possible. Do anything you can to practice under pressure. Eclipse day is not the time to practice.

Once the sun is completely covered, don't just take photographs. Capture the experience and the image of the total solar eclipse in your mind as well. Set up cameras around you to record not just the total solar eclipse but also the excitement and reaction of the crowd.

Eclipse Photography Gear

What do you need to photograph the total eclipse? There are only a few pieces of equipment that you'll need. Preparing to photograph an eclipse successfully takes time. Not only do you have to be skilled and have the right gear, you have to be in the correct place.

Basic Eclipse Photography Equipment

- Solar viewing glasses (verify authenticity)
- Lens solar filter
- Minimum 300mm lens
- Stable tripod that can be tilted to 60° vertical
- High-resolution DSLR
- Spare batteries for everything
- Secondary camera to photograph people, the horizon, etc.
- Remote cable or wireless release

Additional Items

- Video camera
- Video camera tripod
- Quality pair of binoculars
- Solar filters for each binocular lens
- Photo editing software

Equipment to Prepare Before the Big Day

A. Solar viewing glasses

You need a pair of solar viewing glasses as the eclipse approaches.

B. Solar Filter

Partial and total eclipse photography is different from normal photography. Even if only 1% of the sun's surface is visible, it is still approximately 10,000 times brighter than the moon. Before totality, use a solar filter on your lens. Do not look at the sun with your eyes. It can cause irreparable damage to your retinas.

DO NOT leave your camera pointed at the sun without a solar filter attached. The sun will melt the inside of your camera. Think of a magnifying glass used to torch ants and multiply that by one hundred.

C. Lens

To capture the corona's majesty, you need to use a telescope or a telephoto lens. The best focal length, which will give you a large image of the sun's disk, is 400mm and above. You don't want to waste all your efforts by bringing home a small dot where the black disk and majestic corona are supposed to be.

D. Tripod

Bring a stable enough tripod to support your camera properly to avoid unsteady shots and repeated adjustments. Either will ruin your photos. It also needs to be portable in case you need to change locations for a better shot. ***Shut off camera stabilization on a tripod!***

E. Camera

You need to remember to set your camera to its highest resolution to capture all the details. Set your camera to:

- 14-bit RAW is ideal, otherwise
- JPG, Fine compression, Maximum resolution

Bracket your exposures. Shoot at various shutter speeds to capture different brightnesses in the corona. Note that stopping your lens all the way down may not result in the sharpest images.

Choose the lowest possible ISO for the best quality while maintaining a high shutter speed to prevent blurred shots. Set your camera to manual. Do not use AUTO ISO. Your camera will be fooled. The night before, test the focus position of your lens using a bright star or the moon.

Constantly double-check your focus. Be paranoid about this. You can deal with a grainy picture. No amount of Photoshop will fix a blurry, out-of-focus picture.

F. Batteries

Remember to bring fresh batteries! Make sure that you have enough power to capture the most important moments. Swap in fresh batteries thirty minutes before totality.

G. Remote release

Use a wired or wireless remote release to fire the camera's shutter. This will reduce the amount of camera vibration.

H. Video Camera

Run a video camera of yourself. Capture all the things you say and do during the totality. You'll be amazed at your reaction.

I. Photo editing software

You will need quality photo editing software to process your eclipse images. Adobe Lightroom and Photoshop are excellent programs to extract the most out of your images. Become well versed in how to use them at least a month before the eclipse.

J. Smartphone applications

The following smartphone applications will aid in your photography planning: Wunderground, Skyview, Photographer's Ephemeris, Sunrise and Sunset Calculator, SunCalc, and Sun Surveyor among others.

Camera Phones

Smartphone cameras are useful for many things but not eclipse photography. An iPhone 6 camera has a 63° horizontal field of view and is 3264 pixels across. If you attempt to photograph the eclipse, the sun will be a measly 30-40 pixels wide depending on the phone. Digital pinch zoom won't help here. If you want *National Geographic* images, you'll need a serious camera and lens, far beyond any smartphone.

Consider instead using a smartphone to run a time-lapse of the entire event. The sun will be minuscule when shot on a smartphone. Think of something else exciting and interesting do to with it. Purchase a Gorilla Pod, inexpensive tripod, or selfie stick and mount the smartphone somewhere unique.

Also, partial and total eclipse light is strange and ethereal. Consider using that light to take unique pictures of things and people. It's rare and you may have something no one else does.

Focal Length & the Size of Sun

The size of the sun in a photo depends on the lens focal length. A 300mm lens is the recommended minimum on a full-frame (FF) DSLR. Lenses up to this size are relatively inexpensive. For more magnification, use an APS-C (crop) size sensor. Cameras with these sensors provide an advantage by capturing a larger sun.

For the same focal length, an APS-C sensor will provide a greater apparent magnification of any object. As a consequence, a shorter, less expensive lens can be used to capture the same size sun.

The below figure shows the size of the sun on a camera sensor at various focal lengths. As can be seen with the 200mm lens, the sun is quite small. On a full-frame camera at 200mm, the sun will be 371 pixels wide on a Nikon D810, a 36-megapixel body. A lower resolution FF camera will result in an even smaller sun.

Printing a 24-inch image shot on a Nikon D810 with a 200mm lens at a standard 300 pixels per inch results in a small sun. On this size paper, the sun will be a miserly 1.25 inches wide!

Photographing the eclipse with a lens shorter than 300mm will leave you with little to work with. Using a 400mm lens and printing a 24-inch print will result in a 2.5-inch-wide sun. For as massive as the sun is, it is a challenge to take a large photograph of the sun. The sun will appear to move fast with a 500mm lens, too. Plan to adjust.

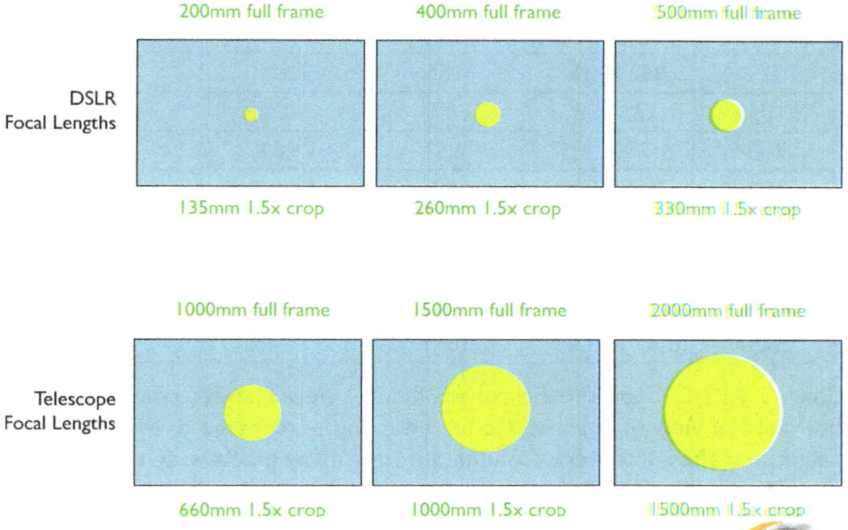

FOCAL LENGTH	FOV FULL FRAME	FF VERT. ANGLE	% OF FF	SUN PIXEL SIZE
14	104° X 81°	81°	0.7%	32.1
20	84° X 62°	62°	0.9%	41.9
28	65° X 46°	46°	1.2%	56.5
35	54° X 38°	38°	1.4%	68.5
50	40° X 27°	27°	2.0%	96.4
105	19° X 13°	13°	4.1%	200.2
200	10° X 7°	7°	7.6%	371.9
400	5° X 3.4°	3.4°	15.6%	765.6
500	4° X 2.7°	2.7°	19.6%	964.2
1000	2° X 1.3°	1.3°	40.8%	2002.5
1500	1.4° X 0.9°	0.9°	58.9%	2892.6
2000	1° X 0.68°	0.68°	77.9%	3828.4

Chart 1: Full-frame camera field of view. The 3rd column is the vertical field of view in degrees. Column 4 is the percentage of the total sensor height that the sun covers. Column 5 is how many pixels wide the sun will be on a 36MP Nikon D810. (Values are estimates)

FOCAL LENGTH	FOV CROP	CROP VERT DEG	% OF CROP	SUN PIXEL SIZE
14	80° X 58°	58°	0.9%	33.9
20	61° X 43°	43°	1.2%	45.8
28	45° X 31°	31°	1.7%	63.5
35	37° X 25°	25°	2.1%	78.7
50	26° X 18°	18°	2.9%	109.3
105	13° X 8°	8°	6.6%	245.9
200	6.7° X 4.5°	4.5°	11.8%	437.2
400	3.4° X 2°	2°	26.5%	983.7
500	2.7° X 1.8	1.8°	29.4%	1093.0
1000	1.3° X 0.9°	0.9°	58.9%	2186.0
1500	0.9° X 0.6°	0.6°	88.3%	3278.9
2000	0.6° X 0.45°	0.5°	117.8%	4371.9

Chart 2: APS-C Crop sensor camera field of view. The 3rd column is the vertical field of view in degrees. Column 4 is the percentage of the total sensor height that the sun covers. Column 5 is how many pixels wide the sun will be on a 12mp Nikon D300s. (Values are estimates)

The big challenge is the cost of the lens. Lenses longer than 300mm are expensive. They also require heavier tripods and specialized tripod heads. The 70-300mm lenses from Nikon, Canon, Tamron, and others are relatively affordable options. It is worth spending time at a local camera shop to try different lenses. Long focal-length lenses are a significant investment, especially for a single event.

To achieve a large eclipse image, you will need a long focal-length lens, ideally at least 400mm. A standard 70-300mm lens set to 300mm will show a small sun. At 500mm, the sun image becomes larger and covers more of the sensor area. The corona will take up a significant portion of the frame. By 1000mm, the corona will exceed the capture area on a full-frame sensor. See the picture in this chapter for sun size simulations for different focal lengths.

Suggested Exposures

To photograph the partial eclipse, the camera must have a solar filter attached. If not, the intense light from the sun may damage (fry) the inside of your camera. This has happened to the author. The exposure depends on the density (darkness) of the solar filter used.

As a starting point, set the camera to ISO 100, f/8, and with the solar filter on, try an exposure of 1/2000. Make adjustments based on the filter used, histogram, and highlight warning.

Turn on the highlight warning in your camera. This feature is commonly called "blinkies." This warning will help you detect if the image is overexposed or not.

Once the Baily's Beads, prominences, and corona become visible, there will only be a few minutes to take bracketed shots. It will take at least eleven shots to capture the various areas of the sun's corona and stars. The brightness varies considerably. No commercially available camera can capture the incredible dynamic range of the different portions of the delicate corona. This requires taking multiple photographs and digitally combining them afterward.

During totality, try these exposure times at ISO 100 and f/8:
1/4000, 1/2000, 1/1000, 1/250, 1/60, 1/30, 1/15, 1/4, 1/2, 1 sec, and 4 sec.

Disable camera/lens stabilization on a tripod!

Photography Time

Set the camera to full-stop adjustments. It will reduce the time spent fiddling. As an example, the author tried the above shot sequence, adjusting the shutter speed as fast as possible.

It took thirty-three seconds to shoot the above 11 shots using 1/3-stop increments. This was without adjusting composition, focus, or anything else but the shutter speed. When the camera was set to full stop increments, it only took twenty-two seconds to step through the same shutter speed sequence. Use a remote release to reduce camera shake.

Assuming the totality lasts less than two minutes, only four shot sequences could be made using 1/3-stop increments. Yet six shot sequences could be made when the camera was set to full stop steps. Zero time was spent looking at the back LCD to analyze highlights and the histogram.

Now add in the bare minimum time to check the highlight warning. It took sixty-three seconds to shoot and check each image using full stops. And that was without changing the composition to allow for sun movement, bumping the tripod, etc. Looking at the LCD ("chimping") consumed **half** of the totality time.

This test was done in the comfort of home under no pressure. In real world conditions, it may be possible to successfully shoot only one sequence. If you plan to capture the entire dynamic range of the totality, you must practice the sequence until you have it down cold. If you normally fumble with your camera, do not underestimate the difficulty, frustration, and stress of total eclipse photography.

Most importantly, trying to shoot this sequence allowed for zero time to simply look at the totality to enjoy the spectacle.

Avoid Last Minute Purchases

You should purchase whatever you think you'll need to photograph the eclipse early. This event will be nothing short of massive. Remember the hot toy of the year? Multiply that frenzy by a thousand. Everyone will want to try to capture their own photo.

Do not wait until the last few weeks before the eclipse to purchase cameras, lenses, filters, tripods, viewing glasses, and associated material. Consider that the totality of the eclipse will streak across

America. Everyone who wants to photograph the eclipse will order at the same time. If you wait until too late to buy what you need, it's conceivable that solar filters to create a total eclipse photo will be sold out in the United States. All filters sold out during the 2017 total eclipse. Whether this happens or not, do not wait to make your purchases. It may be too late.

Practice

You will need to practice with your equipment. Things may go wrong that you don't anticipate. If you've never photographed a partial or total eclipse, taking quality shots is more difficult than you think. Practice shooting the sequence with a midday sun. This will tell you if you have your exposures and timing correct. Figure out what you need well in advance.

Practice photographing the full moon and stars at night. Capture the moon in full daylight to learn how your camera reacts. Astrophotography is challenging and requires practice.

The August 21, 2017, eclipse as seen in Jackson, WY, shot with a Nikon D800 with an 80-400mm lens set to 340mm. The sun is 644 pixels wide on the 7360x4912 image.

This image is shown straight out of the camera without modification. Even with a high-quality camera and lens, photographing an eclipse is challenging.

The eclipse will follow this approximate path on the afternoon of April 8, 2024. Image of downtown Oklahoma City, OK.

Note that this image is a simulation and approximation the sun's path and where the total eclipse may appear from one perspective. Refer to the eclipse position data for a more accurate location.

⊙ is the symbol for the sun and first appeared in Europe during the Renaissance.

☾ is the ancient symbol for the moon.

Viewing Locations Around Oklahoma

Tens of thousands of people will travel to and around Oklahoma to view the total eclipse. There are few obstructions and there is a vast amount of space to view the total eclipse from.

If the weather is questionable, seek out a new location as soon as possible. If you wait until the hour before the eclipse, you may find yourself stuck in traffic, as others will be looking for a better viewing location. Be safe on the roadways, as drivers may be distracted.

This section contains popular, alternative, and little-known locations to watch the eclipse. As long as there are no clouds or smoke from fires, the partial eclipse will be viewable from anywhere in the state.

Suggested Total Eclipse View Points

Towns and Cities

- Antlers
- Big Cedar
- Broken Bow
- Haworth
- Heavener
- Honobia
- Hugo
- Idabel
- Millerton
- Monroe
- Poteau
- Rattan
- Smithville
- Talihina

Oklahoma Total Eclipse Path

Unique Locations

- Broken Bow Lake
- Hugo Lake
- Ouachita National Forest
- Pine Creek State Park
- Raymond Gary State Park
- Red Slough WMA

LOCATIONS

Antlers

Elevation: 512 feet
Population: 2,453
Main road/hwy: US 271

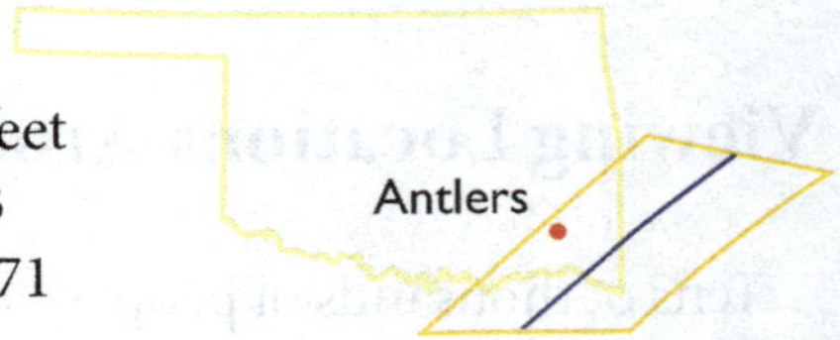

Overview

This city is the county seat of Pushmataha County in the State of Oklahoma. The city has one hotel and two motels, which are Hiway Inn & Suites, Budget Inn, and Sportsman Inn & Suites. Other places of interest here are Wildlife Heritage Center Museum for local wildlife; Rebel Hill Guest Ranch for a comfortable, short-term lodging; Marcos Mexican Restaurant, a historic lunch spot; Ables Western Barbeque; and Cowboy BBQ. Lodging is highly affordable here, and the environment avails you the perfect opportunity to fellowship with nature.

Getting There

Drive west from Hugo on US 271 for 4.7 miles, then turn north and continue on US 271 for another thirteen miles to reach Antlers.

Totality Duration

1 minute 46 seconds

Notes

Visit the Antlers Public Library for updated eclipse information at www.antlerslibrary.okpls.org.

Event	Time (CDT)	Altitude	Azimuth
Sunrise	7:00:00AM	0°	80°
Eclipse Start	12:27:33PM	60°	150°
Totality Start	1:45:43PM	62°	191°
Totality End	1:47:30PM	62°	192°
Eclipse End	3:05:55PM	54°	227°
Sunset	7:48:00PM	0°	279°

Big Cedar

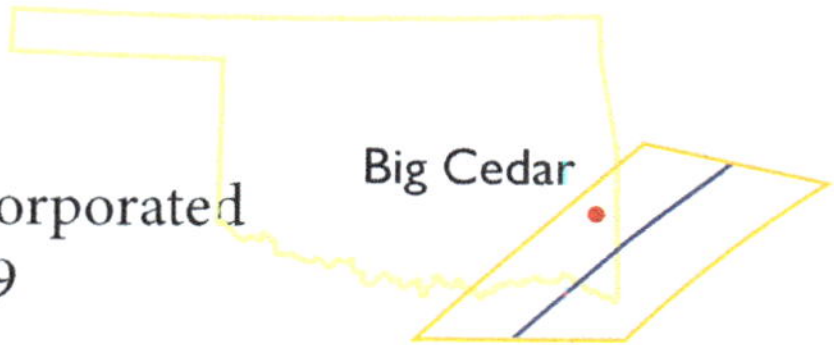

Elevation:	873
Population:	Unincorporated
Main road/hwy:	US 259

Overview

This unincorporated community is under Le Flore County, Oklahoma. The community has several three-star hotels that provide comfortable and inexpensive lodging for tourists, who may be looking for a place to glimpse the 2024 eclipse and an escape from the hustle and bustle of city life. You can spice up your stay here by visiting the north or south shore beach on Cedar Lake. Some of the campgrounds here are Camp Tom Hale, Horse Thief Spring Campground, and Pipe Springs Campground. Big Cedar also features up to five trails for early-morning physical activities or evening hikes.

Getting There

Drive south from Poteau on US 59 past Zoe, then continue south on US 259 to reach Big Cedar.

Totality Duration

3 minutes 21 seconds

Notes

Be aware that the small community of Big Cedar may have limited services during the total eclipse. Please plan accordingly.

Event	Time (CDT)	Altitude	Azimuth
Sunrise	6:55:00AM	0°	80°
Eclipse Start	12:29:43PM	60°	153°
Totality Start	1:47:06PM	62°	194°
Totality End	1:50:27PM	62°	196°
Eclipse End	3:07:50PM	53°	228°
Sunset	7:45:00PM	0°	279°

Broken Bow

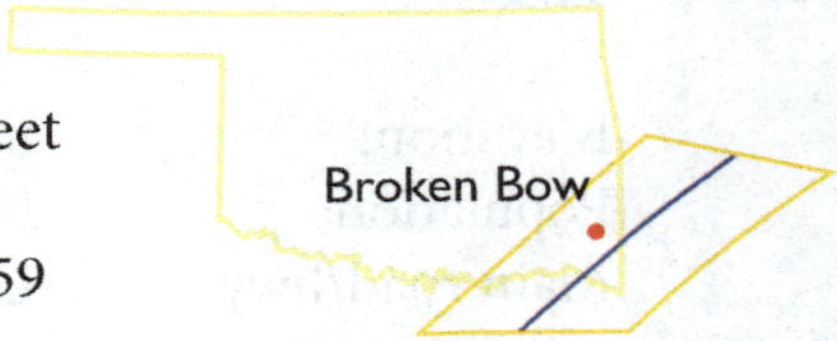

Elevation:	469 feet
Population:	4,120
Main road/hwy:	US 259

Overview

This small city is located at the southeastern end of the state of Oklahoma, and it is popular as a gateway to Beavers Bend State Park, which is the home for pine forest trails, bald eagles, and black bears. One other outstanding spot here is the Forest Heritage Center Museum, which has become a popular spot for tourists. The Broken Bow Lake equally makes this small city attractive to tourists with fishing, boat ramp access, and exquisite beaches. The Hochatown State Park & Petting Zoo, Lower Mountain Fork River, Boat Tours and Water Sports make this place a haven for visitors.

Getting There

Drive east from Idabel on OK 3, then turn north and continue on US 259 for ten miles to reach Broken Bow.

Totality Duration

4 minutes 16 seconds

Notes

The city's website will have eclipse information at www.cityofbrokenbow.com in early 2024.

Event	Time (CDT)	Altitude	Azimuth
Sunrise	6:56:00AM	0°	80°
Eclipse Start	12:28:29PM	60°	152°
Totality Start	1:45:37PM	62°	193°
Totality End	1:49:53PM	62°	196°
Eclipse End	3:07:06PM	53°	229°
Sunset	7:45:00PM	0°	279°

Haworth

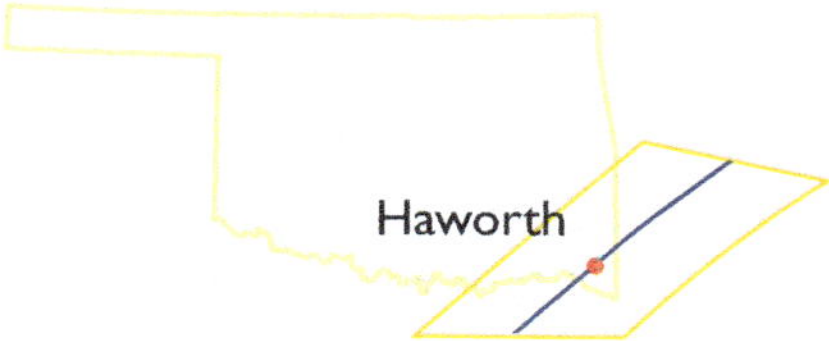

Elevation:	400
Population:	354
Main road/hwy:	OK 3

Overview

Haworth is a town located in McCurtain County, Oklahoma. It covers a total area of 1.6 square miles. This town is landlocked, but it has its fair share of mountains. It may not be a haven for anglers, but hikers can surely have some fun here. Also, the town has several recreation spots and parks to its name; principal among them are Haworth Country Club and White Beeches Golf & Country Club. The multitude of vacation rentals here creates a comfortable home for tourists, enabling them to fellowship with nature while they are here.

Getting There

Drive east on OK 3 from Idabel and continue south on OK-3 until you reach Haworth.

Totality Duration

4 minutes 19 seconds

Notes

Haworth is a small rural community and may have limited resources during the total eclipse.

Event	Time (CDT)	Altitude	Azimuth
Sunrise	6:55:00AM	0°	80°
Eclipse Start	12:28:17PM	61°	152°
Totality Start	1:45:29PM	63°	193°
Totality End	1:49:48PM	62°	196°
Eclipse End	3:07:03PM	54°	229°
Sunset	7:45:00PM	0°	279°

Heavener

Elevation:	554
Population:	3,414
Main road/hwy:	US 59

Overview

This Le Flore County city is part of the Fort-Smith-Arkansas-Oklahoma Metropolitan Statistical Area, and it got its name from Joseph H. Heavener, an 1877 settler in this locality. One of the most notable features of this city is the Heavener Runestones, located just outside its limits. Moreover, the city is an agricultural haven. The Talimena Scenic Byway is the perfect terrain for a cool evening ride. Also, Ed and Linda's 128 Bar and Grill center ensures that your tummy also shares in the fun.

Getting There

Drive south on US 59 from Poteau for thirteen miles to reach Heavener.

Totality Duration

2 minutes 32 seconds

Notes

Visit the Heavener Chamber of Commerce website for eclipse updates at the beginning of 2024 at www.heavenerchamber.com.

Event	Time (CDT)	Altitude	Azimuth
Sunrise	6:54:00AM	0°	80°
Eclipse Start	12:30:14PM	60°	154°
Totality Start	1:47:56PM	62°	194°
Totality End	1:50:29PM	61°	196°
Eclipse End	3:08:08PM	53°	228°
Sunset	7:45:00PM	0°	279°

Honobia

Elevation:	1,625 feet
Population:	170
Main road/hwy:	OK 144

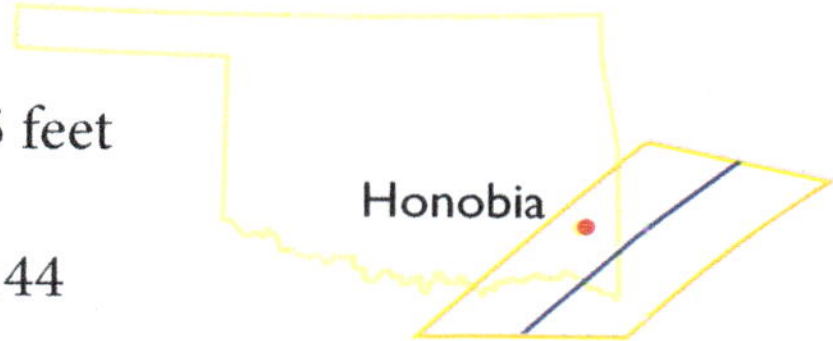

Overview

This unincorporated community is located some fifteen miles from Talihina between Pushmataha County and Le Flore County, both in Oklahoma. The community built a United States Post Office in 1919 and later housed a Choctaw settlement after the Choctaw Indians relocated to the Trail of Tears. This area was initially part of Wade County before statehood. The Honobia Bigfoot Festival is hosted here annually, and all are welcome. Additionally, the Honobia Creek Store & RV Park offers delicious shakes, french fries, fried okra, great burger veggies, and mashed potatoes at highly affordable prices.

Getting There

Drive northeast from Antlers on US 271 and continue on OK 144 for fifty-one miles to reach Honobia.

Totality Duration

2 minutes 57 seconds

Notes

Honobia is a small town and may have limited services during the eclipse. Please plan accordingly.

Event	Time (CDT)	Altitude	Azimuth
Sunrise	6:56:00AM	0°	80°
Eclipse Start	12:29:07PM	60°	153°
Totality Start	1:46:42PM	62°	193°
Totality End	1:49:39PM	62°	195°
Eclipse End	3:07:18PM	53°	228°
Sunset	7:46:00PM	0°	279°

Hugo

Elevation:	545 feet
Population:	5,310
Main road/hwy:	US 271

Overview

Hugo Lake is one of the natural features in this city. The city is the county seat of Choctaw County, Oklahoma, named after Victor Hugo, a French novelist, in 1901, and it also serves as the winter quarters for many circus performers. Goodland Academy is one of the unique tourist attractions in this city. The school is one of the oldest on the western side of Mississippi, adjacent to the city. The Hugo Lake State Park, Frisco Depot Museum, Showman's Rest and Bull Rider's Reprieve, and the Endangered Ark Foundation make this city a haven for tourism.

Getting There

Drive south and then east on US 271 for twenty-one miles from Antlers to reach Hugo.

Totality Duration

3 minutes 18 seconds

Notes

The Hugo website, www.hugook.com, will have eclipse updates in early 2024 regarding lodging and visitation.

Event	Time (CDT)	Altitude	Azimuth
Sunrise	6:59:00AM	0°	80°
Eclipse Start	12:27:20PM	60°	150°
Totality Start	1:44:51PM	63°	191°
Totality End	1:48:09PM	63°	193°
Eclipse End	3:05:54PM	54°	227°
Sunset	7:48:00PM	0°	279°

Idabel

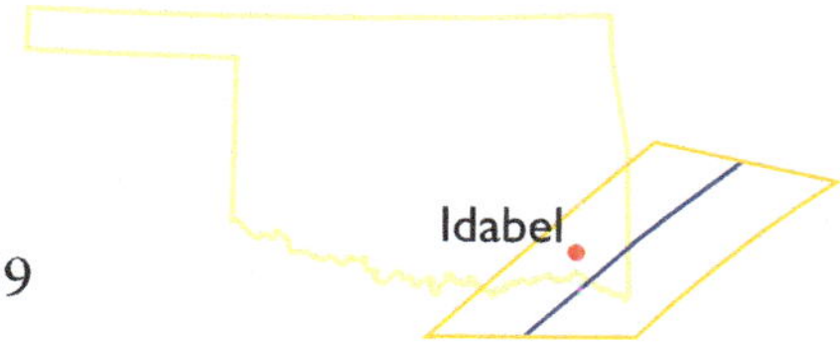

Elevation:	472
Population:	7,010
Main road/hwy:	US 259

Overview

This city is the county seat of McCurtain County, Oklahoma. It is a very popular tourist area called Kiamichi Country and lies between the Red River and the Little River. The city was birthed in 1902 by the Choctaw and Arkansas Railway. Some of the points of interest in this city are Little River National Wildlife Refuge, Choctaw Casino, and Museum of the Red River. The Comfort Suites, Super 8, and Idabel Inn are few of the hotels here; they offer comfortable, affordable, and homey lodging for tourists.

Getting There

Idabel is located in the southeast corner of Oklahoma and is the starting point for many towns in this guide. Drive west from Texarkana (TX, AR) on US 82, then turn north on US 259 once you pass the town of De Kalb.

Totality Duration

4 minutes 18 seconds

Notes

Visit the city's official website at www.idabel-ok.gov for eclipse, visitation, and lodging updates.

Event	Time (CDT)	Altitude	Azimuth
Sunrise	6:56:00AM	0°	80°
Eclipse Start	12:28:07PM	61°	152°
Totality Start	1:45:17PM	63°	193°
Totality End	1:49:35PM	62°	195°
Eclipse End	3:06:50PM	54°	229°
Sunset	7:45:00PM	0°	279°

Millerton

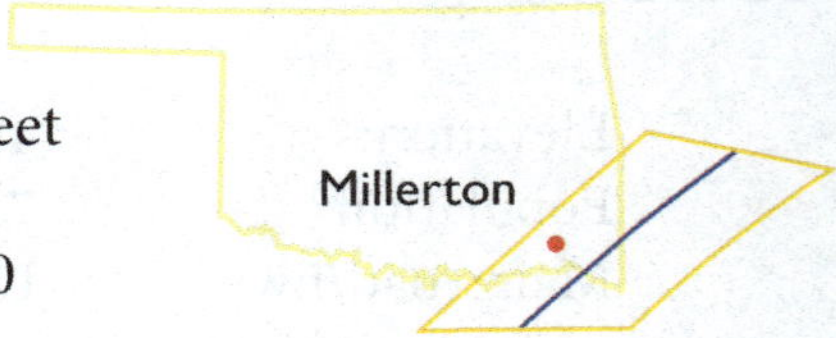

Elevation:	505 feet
Population:	359
Main road/hwy:	US 70

Overview

This village is located in McCurtain County, Oklahoma. Several individuals contributed to the growth of this town; one of them is Jeromy Mixon. He once donated up to $250,000 to the town, which helped with the Fire Protection and Civil Improvement Projects of the village. While you are here, remember to visit the Wheelock Academy located on Locust Street in Millerton. It was established for Native American girls and stopped functioning in the mid-1950s. Some of the buildings on its premises have been converted to museums.

Getting There

Drive northwest from Idabel for fourteen miles to reach Millerton.

Totality Duration

4 minutes 6 seconds

Notes

The community of Millerton may have limited services during the total eclipse. Please plan accordingly.

Event	Time (CDT)	Altitude	Azimuth
Sunrise	6:57:00AM	0°	80°
Eclipse Start	12:28:00PM	60°	151°
Totality Start	1:45:11PM	63°	192°
Totality End	1:49:18PM	62°	195°
Eclipse End	3:06:38PM	54°	228°
Sunset	7:46:00PM	0°	279°

Monroe

Elevation:	72 feet
Population:	531
Main road/hwy:	OK 83

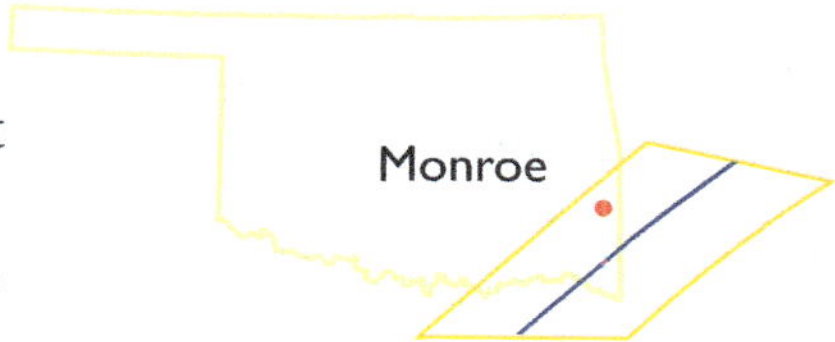

Overview

This Oklahoma town is a small, unincorporated community located in the eastern part of the Le Flore County in Oklahoma. The town has quite a history behind it. For example, its post office was built in 1881. The town got its name from the first postmaster to work here, Simon Monroe Griffith. While you are here, remember to attend the annual event Lights on the Island. It started about ten years ago but now contributes significantly to tourism in this town. Participants come from other towns in Oklahoma and far beyond.

Getting There

Drive south on US 59 from Poteau, then continue east on OK 83 to reach the community of Monroe.

Totality Duration

2 minutes 18 seconds

Notes

The community of Monroe has limited visitor services. Plan ahead with fuel, water, and food during the rush of the eclipse.

Event	Time (CDT)	Altitude	Azimuth
Sunrise	6:55:00AM	0°	80°
Eclipse Start	12:30:31PM	60°	154°
Totality Start	1:48:19PM	61°	195°
Totality End	1:50:37PM	61°	196°
Eclipse End	3:08:20PM	53°	228°
Sunset	7:45:00PM	0°	279°

Poteau

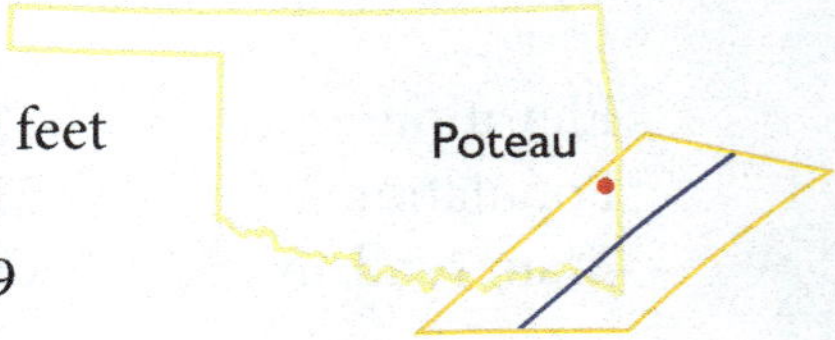

Elevation:	1,999 feet
Population:	7,939
Main road/hwy:	US 59

Overview

Poteau city is the seat of Le Flore County. It was founded in 1885 and got its name from the Poteau River, which was originally named by a group of French explorers in the early eighteenth century. The city has one of the highest temperature recordings in the state of Oklahoma. One of the areas of interest is Poteau Convention & Visitors Bureau, where you can get information about points of interest in this city. Other places you should visit while you are here are the Poteau Farmer's Market and Bazar, Tower Drive-In, and LeFlore County Museum.

Getting There

Drive southeast from Tulsa, OK, on OK 351, turn east on I-40, then continue south on US 59 for one hundred twenty-eight miles to reach the city of Poteau.

Totality Duration

1 minute 19 seconds

Notes

Contact the Poteau Chamber of Commerce for lodging, event planning, and other eclipse activities at poteauchamber.com.

Event	Time (CDT)	Altitude	Azimuth
Sunrise	6:54:00AM	0°	80°
Eclipse Start	12:30:29PM	60°	154°
Totality Start	1:48:44PM	61°	195°
Totality End	1:50:04PM	61°	195°
Eclipse End	3:08:15PM	53°	228°
Sunset	7:45:00PM	0°	279°

Rattan

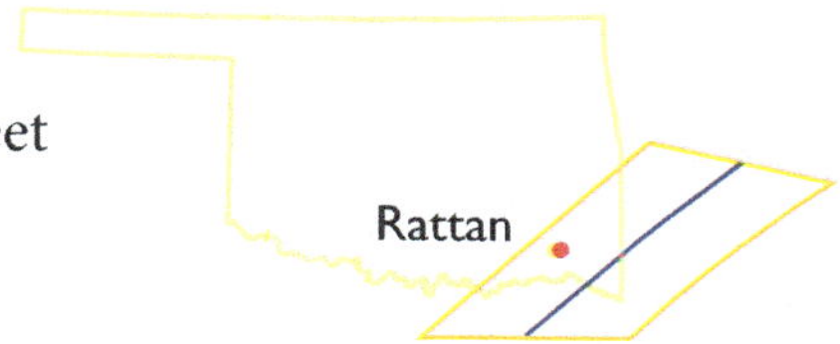

Elevation:	466 feet
Population:	241
Main road/hwy:	OK 3

Overview

This town in Pushmataha County was founded in 1910. The natural resources found here contributed a great deal to its early growth. One such natural resource is timber, found in the Kiamichi Mountains, towards the northern part of the town. Many of those who settled here at its inception were farmers. The community has general stores, physician offices, small schools, and everything that can make a small town function effectively. While you are here, visit the Choctaw Casino Resort for top-class events that are held in this venue several times every year.

Getting There

Drive east from Antlers for 12.5 miles on OK 3 to reach Rattan.

Totality Duration

2 minutes 51 seconds

Notes

The town of Rattan has limited services. Contact the city hall in Rattan for eclipse information at (580) 587-2256.

Event	Time (CDT)	Altitude	Azimuth
Sunrise	6:58:00AM	0°	80°
Eclipse Start	12:27:48PM	60°	151°
Totality Start	1:45:29PM	62°	192°
Totality End	1:48:21PM	62°	193°
Eclipse End	3:06:14PM	54°	227°
Sunset	7:48:00PM	0°	279°

Talihina

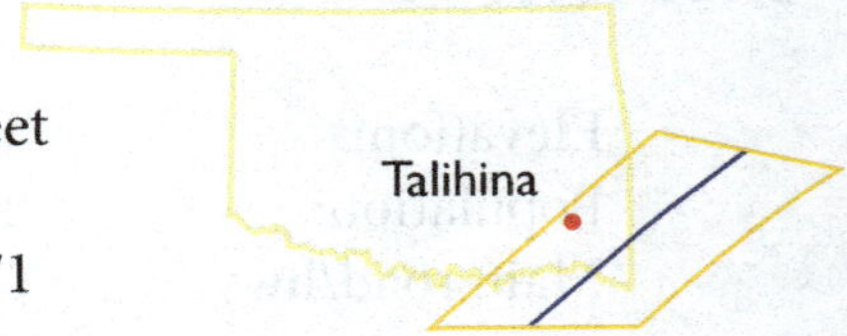

Elevation:	682 feet
Population:	1,114
Main road/hwy:	US 271

Overview

The town of Talihina was originally named by the Choctaw Indians in 1886 for the railroad that ran through the area. Over the time that the town has been in existence, six reservoirs have been built in the Kiamichi Valley, attracting outdoor and watersports enthusiasts for decades. There is the Talimena State Park and the Old Military Road for visitors to enjoy before and after the eclipse. Talihina is also the western terminus of the Talimena Scenic Drive, starting in Mena, Arkansas.

Getting There

Drive southwest from Poteau on US 271 for thirty-eight miles to reach the town of Talihina.

Totality Duration

1 minute 4 seconds

Notes

Visit Talihina's Chamber of Commerce for eclipse updates at www.talihinacc.com.

Event	Time (CDT)	Altitude	Azimuth
Sunrise	6:57:00AM	0°	80°
Eclipse Start	12:29:20PM	60°	153°
Totality Start	1:47:46PM	62°	193°
Totality End	1:48:50PM	62°	194°
Eclipse End	3:07:20PM	53°	228°
Sunset	7:46:00PM	0°	279°

BROKEN BOW LAKE

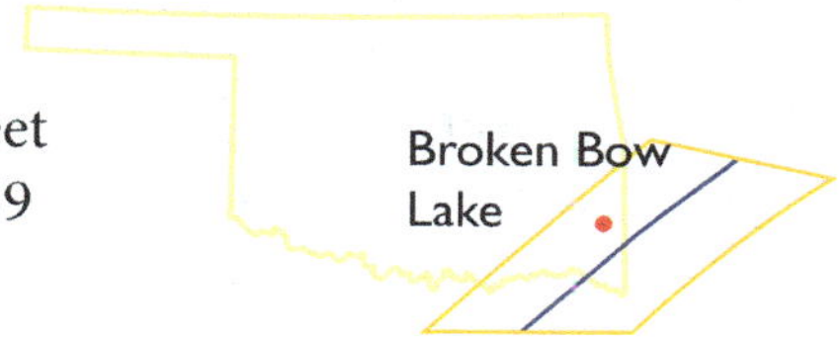

Elevation: 601 feet
Main road/hwy: US 259

OVERVIEW

The lake is a reservoir located in the southeastern part of Oklahoma, close to Mountain Fork River and about nine miles from Broken Bow, a city in McCurtain County. It is unique in that the bottom is covered in rocks rather than mud, giving the water an incredible transparent clarity. This lake is among the largest lakes in Oklahoma and is a popular destination for locals as well as tourists. The lake covers an expanse of twenty-two square miles, with its end set into the Ouachita Mountain. The Beaver Lodge Nature Trail and Big Oak Nature Trail make this place the perfect fun spot for hikers. Small children and senior citizens will, however, find the Big Oak trail more welcoming than the former.

GETTING THERE

Drive north from Idabel on US 259, then turn east on E1965 near Hochatown to reach Broken Bow Lake.

TOTALITY DURATION

4 minutes 9 seconds

NOTES

visitmccurtaincounty.com/parks/parks-rivers-lakes/broken-bow-lake

Event	Time (CDT)	Altitude	Azimuth
Sunrise	6:55:00AM	0°	80°
Eclipse Start	12:28:52PM	60°	153°
Totality Start	1:46:01PM	62°	194°
Totality End	1:50:11PM	62°	196°
Eclipse End	3:07:21PM	53°	229°
Sunset	7:45:00PM	0°	279°

Hugo Lake State Park

Elevation:	404 feet
Main road/hwy:	US 70

Overview

This is a seven-mile man-made lake located east of Hugo, Choctaw County. The Hugo Lake originates from the Hugo Lake Dam located on Kiamichi River, which is located some eighteen miles from Red River. This lake is the perfect spot for various watersports, like boating and water skiing. Some of the available facilities include cabins, hiking trails, and a fifty-six-slip marina. Anglers are provided with an ideal condition for fishing thanks to the stumps of trees allowing cover for fishes in the north end of this lake.

Getting There

Drive south from Antlers on US 271, then continue on US 70 until you reach Fallon. Drive north to the park. The total driving distance is thirty-one miles.

Totality Duration

3 minutes 22 seconds

Notes

Visit the park's website at www.hugolakestatepark.com to make reservations well in advance of the total eclipse.

Event	Time (CDT)	Altitude	Azimuth
Sunrise	6:59:00AM	0°	80°
Eclipse Start	1:27:34PM	60°	151°
Totality Start	1:45:02PM	62°	192°
Totality End	1:48:25PM	62°	193°
Eclipse End	3:06:06PM	54°	227°
Sunset	7:47:00PM	0°	279°

Ouachita National Forest

Elevation:	Various
Main road/hwy:	Multiple

Overview

This forest is managed by the US Forest Service, and it is one of the greatest places in Oklahoma for camping. It is located close to Lake Ouachita, which has one thousand miles of shoreline. The forest is huge and covers a space of 1.7 million acres. It is also dotted by mountainous terrains making it the perfect location for mountain hiking. Also, the forest was established in 1907, and it is the oldest forest in the southern part of the United States. Some of the recreation spots and activities for tourists at this location are equestrian trails, all-terrain vehicle trails, mountain bike trails, fishing, hunting, and swimming.

Getting There

Use the town of Big Cedar as your starting point for the forest.

Totality Duration

3 minutes 15 seconds (for Page)

Notes

Contact the forest through their website for eclipse updates at www.fs.usda.gov/ouachita.

Times for Page

Event	Time (CDT)	Altitude	Azimuth
Sunrise	6:55:00AM	0°	80°
Eclipse Start	12:29:55PM	60°	154°
Totality Start	1:47:20PM	62°	194°
Totality End	1:50:36PM	62°	196°
Eclipse End	3:07:59PM	53°	228°
Sunset	7:44:00PM	0°	279°

Pine Creek State Park

Elevation: 530
Main road/hwy: OK 3

Overview

This is yet another unique park just perfect for camping and other outdoor adventures. The pine lake shoreline spreads for seventy-four miles, making it an exquisite location for seasoned outdoor enthusiasts who may have a thing for fishing. Fishing enthusiasts can play around here with sunfish, crappie, and bass to keep them company. You can take a ride on the lake by launching from any of the nearby boat ramps. Licensed hunters can enjoy the game management area that covers a ten-thousand-acre expanse of land. Aside from being a perfect hunting ground, this location is also excellent for picnicking.

Getting There

Drive west from Antlers on OK 3 for twenty-five miles, then follow these roads to the state park: Co. Rd. 4410, Co. Rd. 1930, E1930, and then N4430.

Totality Duration

3 minutes 27 seconds

Notes

www.stateparks.com/pine_creek_state_park_in_oklahoma.html

Event	Time (CDT)	Altitude	Azimuth
Sunrise	6:58:00AM	0°	80°
Eclipse Start	12:28:15PM	60°	151°
Totality Start	1:45:40PM	62°	192°
Totality End	1:49:07PM	62°	194°
Eclipse End	3:06:41PM	54°	228°
Sunset	7:46:00PM	0°	279°

Raymond Gary State Park

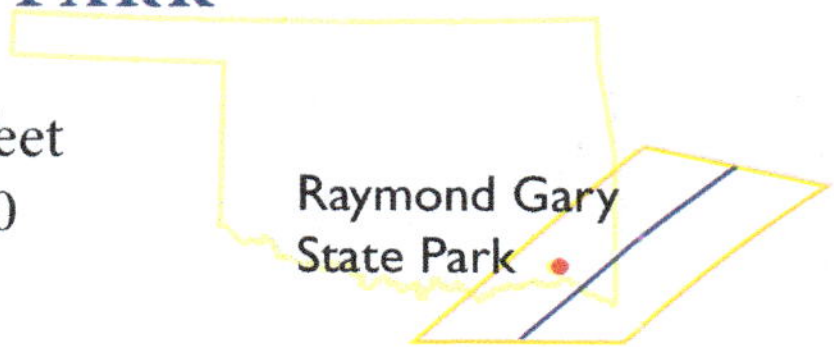

Elevation: 440 feet
Main road/hwy: US 70

Overview

Raymond Gary State Park is located on the shore of Lake Raymond Gary on the southeastern end of Oklahoma. Various activities like swimming, boating, fishing, picnicking, and camping take place here. While here, tourists can access up to one hundred tent sites, nine sites with water and electricity hookups, and ten RV sites with full hookups. Also, twin bed, heat- and air-control system, microwave, mini-refrigerator, and fireplace are all amenities in each of the six primitive lakefront cabins open to tourists here. Additionally, lodgers are provided with toiletries, towels, and linens. Accommodations are affordable and of top quality. Visitors are always treated to enjoyable experiences here.

Getting There

Drive east from Hugo on US 70 for seventeen miles to reach the park.

Totality Duration

3 minutes 47 seconds

Notes

www.stateparks.com/raymond_gary_state_park_in_oklahoma.html

Event	Time (CDT)	Altitude	Azimuth
Sunrise	6:58:00AM	0°	80°
Eclipse Start	12:27:40PM	60°	151°
Totality Start	1:44:59PM	63°	192°
Totality End	1:48:46PM	62°	194°
Eclipse End	3:06:16PM	54°	228°
Sunset	7:47:00PM	0°	279°

Red Slough Wildlife Management Area

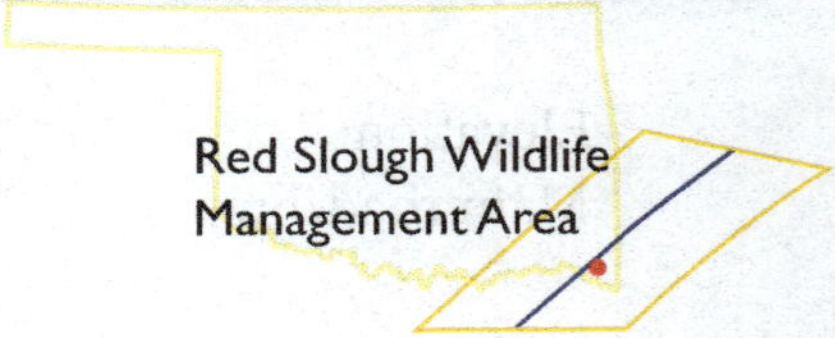

Elevation:	336
Main road/hwy:	OK 87

Overview

Red Slough Wildlife Management Area is a wetland project covering a total of 5,814 acres. It is located in the southeastern part of Oklahoma. Additionally, it is a premier waterfowl hunting and bird-watching area. Its purpose is to reestablish bottomland hardwoods and restore hydrology. This project is being managed by the Oklahoma Department of Wildlife Conservation, Natural Resources Conservation Service, and the US Forest Service. The objective of this project is to offer outstanding recreational opportunities to tourists while maximizing nature's diversity. This location is a unique hideaway for anyone seeking to escape the hustle and bustle of city life and also the perfect place to view the 2024 eclipse.

Getting There

Drive south from Idabel on US 259 and continue on OK 87 to reach the management area's entrance.

Totality Duration

4 minutes 18 seconds

Notes

www.wildlifedepartment.com/wildlife-management-areas/red-slough

Event	Time (CDT)	Altitude	Azimuth
Sunrise	6:56:00AM	0°	80°
Eclipse Start	12:28:18PM	61°	152°
Totality Start	1:45:30PM	63°	194°
Totality End	1:49:49PM	62°	196°
Eclipse End	8:07:05PM	54°	229°
Sunset	7:44:00PM	0°	279°

Smithville Park

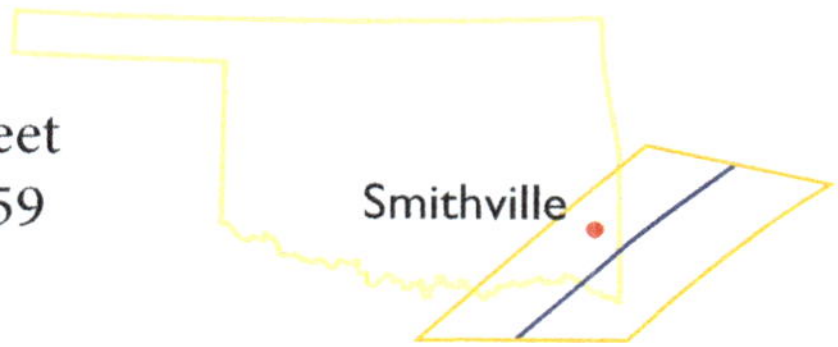

Elevation: 721 feet
Main road/hwy: US 259

Overview

This historic park is the centerpiece of the McCurtain County Parks system. Smithville Park features a visitors center, playground for kids, trails for an evening walk or early morning jog, picnic area for the family and loved ones, fishing areas, a floating walkway, and a butterfly garden. Additionally, make sure to visit the historical museums, mansions, and galleries in the park when you are here for the eclipse. Furthermore, tourists will have access to various amenities, like restrooms, Wi-Fi, and charcoal grills. The picnic areas can accommodate up to fifty people at one time.

Getting There

Drive north on US 259 from Idabel for fifty-one miles to reach Smithville Park.

Totality Duration

3 minutes 48 seconds

Notes

Make your reservations early if you plan to have an event at the park. Visit the Boondockers Landing River Resort for event information at www.boondockerslanding.com.

Event	Time (CDT)	Altitude	Azimuth
Sunrise	6:56:00AM	0°	80°
Eclipse Start	12:29:24PM	60°	153°
Totality Start	1:46:38PM	62°	194°
Totality End	1:50:26PM	62°	196°
Eclipse End	3:07:40PM	53°	228°
Sunset	7:45:00PM	0°	279°

Paste a meaningful picture of your eclipse experience, perhaps a photo of you and the people you watched it with. This will help you remember your family, friends, and companions experience together.

Remember the Oklahoma Total Eclipse

April 8, 2024

Who was I with? ______________________________

____

____

____

____

What did I see? ______________________________

____

____

____

____

____

____

____

____

What did I feel? ______________________________

____

____

____

____

____

____

____

What did the people with me think?

Where did I stay?

Enjoy Other Books by Aaron Linsdau

50 Jackson Hole Photography Hotspots

This guide reveals the best Jackson Hole photography spots. Learn what locals and insiders know to find the most impressive and iconic photography locations in the United States. This is an excellent companion guide to the *Jackson Hole Hiking Guide*.
www.sastrugipress.com/books/50-jackson-hole-photography-hotspots/

Adventure Expedition One
by Aaron Linsdau M.S. & Terry Williams, M.D.

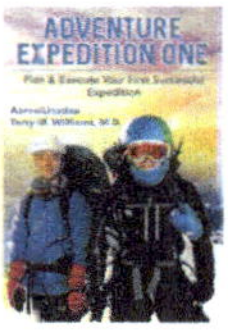

Create, finance, enjoy, and return safely from your first expedition. Learn the techniques explorers use to achieve their goals and have a good time doing it. Acquire the skills, find the equipment, learn to camp, understand medical issues, and learn the planning necessary to pull off an expedition.
www.sastrugipress.com/books/adventure-expedition-one/

Antarctic Tears

Experience the honest story of solo polar exploration. This inspirational true book will make readers both cheer and cry. Coughing up blood and fighting skin-freezing temperatures were only a few of the perils Aaron Linsdau faced. Travel with him on a world-record expedition to the South Pole.
www.sastrugipress.com/books/antarctic-tears/

How to Keep Your Feet Warm in the Cold

Keep your feet warm in cold conditions on chilly adventures with techniques described in this book. Packed with dozens and dozens of ideas, learn how to avoid having cold feet ever again in your outdoor pursuits.
www.sastrugipress.com/books/how-to-keep-your-feet-warm-in-the-cold/

Jackson Hole Hiking Guide

Jackson Hole contains some of the most dramatic and iconic landscapes in the United States. The book shares everything you need to know to hike Jackson's classic trails with canyons, high mountains, and hidden alpine lakes. This book is an excellent companion guide to *50 Jackson Hole Photography Hotspots.*
www.sastrugipress.com/books/jackson-hole-hiking-guide/

Subscribe to Aaron's YouTube channel at www.youtube.com/@alinsdau

If you enjoyed this book, please consider leaving a five-star review and a few words on what you liked about it at your favorite online retailer.

Lost at Windy Corner

Windy Corner on Denali has claimed fingers, toes, and even lives. What would make someone brave lethal weather, crevasses, and avalanches to attempt to summit North America's highest mountain? Aaron Linsdau shares the experience of climbing Denali alone and how you can apply the lessons to your life.
www.sastrugipress.com/books/lost-windy-corner/

The Motivated Amateur's Guide to Winter Camping

Winter camping is one of the most satisfying ways to experience the wilderness. It is also the most challenging style of overnighting in the outdoors. Learn 100+ tips from a professional polar explorer on how to winter camp safely and be comfortable in the cold.
www.sastrugipress.com/books/the-motivated-amateurs-guide-to-winter-camping/

Two Friends and a Polar Bear
by Terry Williams, M.D. & Aaron Linsdau

This story of friendship is about two old friends who plan to ski across the Greenland Ice Cap along the Arctic Circle in hopes of becoming one of the oldest teams to succeed.
www.sastrugipress.com/books/two-friends-and-a-polar-bear/

About the Author

Aaron Linsdau is the second American to ski alone from the coast of Antarctica to the South Pole (730 miles / 1174 km), setting a world record for surviving the longest expedition ever for that trip. He lead a 310-mile (499 km) ski expedition across the Greenland icecap along the Arctic Circle. Aaron has climbed Denali solo, crossed the Greenland tundra alone, skied across Yellowstone National Park solo, trekked through the Sahara desert, and successfully climbed Mt. Kilimanjaro and Mt. Elbrus in Russia.

Aaron Linsdau at the South Pole.

Use your smart device to scan the QR codes for website links.

Visit www.aaronlinsdau.com/subscribe to learn more about the author. Receive updates when he releases new books and shows.

Visit Sastrugi Press on the web at www.sastrugipress.com to purchase the above titles in bulk. They are available in print, e-book, or audiobook form.

Thank you for choosing Sastrugi Press.

Enjoy Other Books by Sastrugi Press

50 Florida Wildlife Hotspots by Moose Henderson Ph.D.

This is a definitive guide to finding where to photograph wildlife in Florida. Follow the guidance of a professional wildlife photographer as he takes you to some of the best places to see wildlife in the Sunshine State.

www.sastrugipress.com/books/50-florida-wildlife-hotspots/

50 Wildlife Hotspots Grand Teton National Park by Moose Henderson Ph.D.

Find out where to find animals and photograph them in Grand Teton National Park from a professional wildlife photographer. Learn techniques, timing, animal behavior, and composition to create stunning wildlife images.

www.sastrugipress.com/books/50-wildlife-hotspots/

Alaska: A Guide for the Curious by Nikki Mann & Jeff Wohl

Discover the natural world of Alaska. Find out what the plants and animals are like, how to identify them, and what the environment of Alaska is like.

www.sastrugipress.com/books/alaska-a-guide-for-the-curious/

Blood Justice by Tim W. James

Two brothers, one a preacher's son, the other an adopted would-be slave, set out in opposite directions to avenge their family's murder only to cross paths in pursuit of the killer. Book 1 of the Roger Brinkman Series.

www.sastrugipress.com/iron-spike-press/blood-justice/

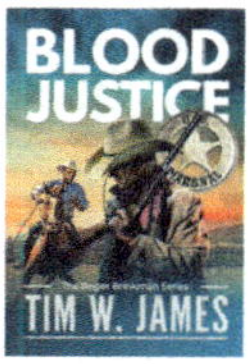

Shake Yourself Free by Bob Millsap

Learn how to overcome difficult encounters with misfortune, tragedy, and loss. Emotional recovery is a journey requiring a mindset shift. Get this book now and take control of your life.

www.sastrugipress.com/books/shake-yourself-free/

The Burqa Cave by Dean Petersen

Still haunted by Iraq, a retired soldier seeks solace teaching high school in Wyoming. He soon finds the quiet town is home to murderers, maniacs, and a boy who can see where missing murder victims are. This paranormal thriller-romance surprises readers with unexpected twists.

www.sastrugipress.com/books/the-burqa-cave/

Photography Notes

April 8, 2024

Settings ______________________________

Successes ______________________________

Challenges ______________________________

Photography Notes

April 8, 2024

Camera

Company

Location